The Institute of Biology's
Studies in Biology no. 2

Life in the Soil

by Richard M. Jackson Ph.D.
Lecturer, Department of Biological and Health Studies, University of Surrey

Frank Raw Ph.D.
Sometime of Rothamsted Experimental Station

Edward Arnold (Publishers) Ltd

First published 1966
Reprinted 1967
Reprinted 1970

SBN: 7131 2080 0 Boards
SBN: 7131 2081 9 Paper

Printed in Great Britain by
William Clowes and Sons Ltd, London and Beccles

General Preface to the Series

It is no longer possible for one textbook to cover the whole field of Biology and to remain sufficiently up to date. At the same time students at school, and indeed those in their first year at universities, must be contemporary in their biological outlook and to know where the most important developments are taking place.

The Biological Education Committee, set up jointly by the Royal Society and the Institute of Biology, is sponsoring, therefore, the production of a series of booklets dealing with limited biological topics in which recent progress has been most rapid and important.

A feature of the series is that the booklets indicate as clearly as possible the methods that have been employed in elucidating the problems with which they deal. There are suggestions for practical work for the student which should form a sound scientific basis for his understanding.

1966 INSTITUTE OF BIOLOGY
41 Queen's Gate London, S.W.7.

Preface

Soil biology is an excellent subject for study in schools and colleges because the soil, with which many of man's activities are concerned, is so readily accessible and contains such a great variety of organisms in a closely integrated ecosystem. Soil biology can also provide a valuable introduction to the study of other ecosystems.

The purpose of this booklet is to introduce the student to the soil as an environment for life, and then to describe the organisms inhabiting it as well as their characteristic activities. The range of soil organisms is so great that only a selection of topics can be discussed in detail in a booklet of this size. Others, sometimes equally important, had to be omitted. We have, however, endeavoured to select topics that are representative of the range of soil organisms and provide examples of their biology, physiology and ecology that can be studied by school and college students. The experimental methods described have been chosen particularly for their suitability for use in an average school or college laboratory.

We are indebted to many of our colleagues at Rothamsted and elsewhere for helpful discussions and suggestions and we would welcome comments, particularly on the experimental methods, from those who use this booklet.

R. M. J. and F. R.

Contents

The Soil Environment 1

1.1 Soil formation

Soil can be defined as the material that plants grow in. It contains various proportions of mineral and organic materials and extends from the ground surface to the lower limit of root growth. It is formed by the interaction of complex processes including the physical and chemical weathering of parent rock material that provides the mineral substrate, the incorporation and decay of organic matter (mainly as plant remains that soil microorganisms decompose) and the movement of soluble or suspended materials in percolating or diffusing water. These processes are themselves conditioned by the persistence of the ground surface and its associated cover of vegetation.

1.2 Soil horizons

The continual interaction of these processes leads gradually to the formation of more or less distinct horizontal layers, termed soil horizons, that differ in their physical, chemical and biological characteristics. Collectively they form the soil profile. In normal mineral soils the profile commonly has three main horizons, usually designated A, B and C, and visibly differentiated by such features as colour, texture, physical structure, porosity, the amount and kind of organic matter, root growth and the microbial and other organisms present (Fig. 1–1).

The A horizon includes the surface layer, or 'top-soil', darkened by incorporation of organic matter, in which root growth and other biological activity is greatest and where, as in fertile grassland and cultivated soils, plant remains are rapidly decomposed and intimately mixed with the mineral fraction. Usually the organic matter does not exceed 15% and consists of finely divided humus intimately mixed with the clay-size mineral particles. Sometimes, as in coniferous forests and moorlands, the organic matter occurs in various stages of decomposition as separate layers (O horizons) at the surface of the mineral soil and, in some situations, may accumulate as peat.

The underlying B horizon, or 'sub-soil', is usually more brightly coloured than the A horizon and differs in structure and composition from the relatively unaltered C horizon beneath it, because of the effects of weathering processes. In well-drained soils the B horizon is commonly coloured brown or reddish by iron oxides and may contain more clay than the C horizon because clay has accumulated as a result of weathering. It may also be enriched by materials such as clay, iron oxides and humus washed down from the A horizon. Then a lighter coloured clay-deficient

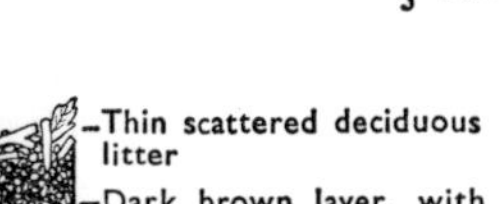

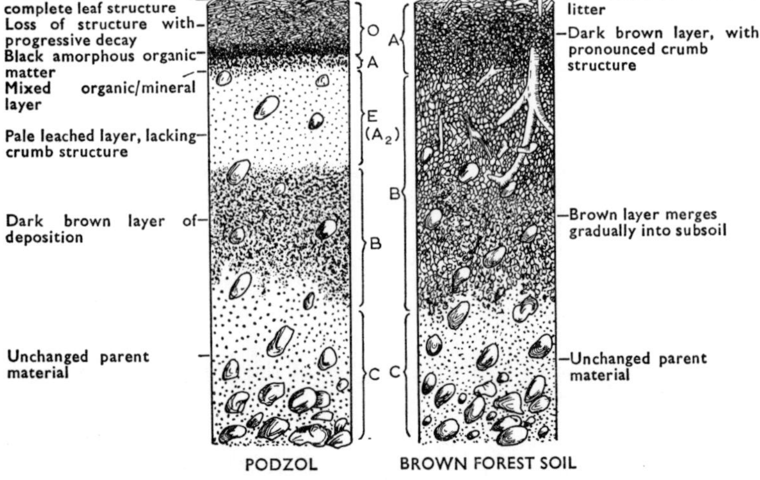

Fig. 1–1 Soil Profiles. (After Fenton, *J. anim. Ecol.* 1947.)

sub-surface horizon, poor in organic matter (designated E or A_2) may occur between the dark coloured A horizon and the B horizon.

In poorly drained soils, as on impermeable clays or in valleys where the water table in winter is near the soil surface, the B horizon may be mottled with grey and ochreous colours attributable to alternate reduction, mobilization and re-oxidation of iron compounds, because of seasonal water-logging.

There is always an A horizon wherever the ground is covered with vegetation, but shallow soils on hard, little-weathered rock and soils formed recently in unconsolidated sediments such as dune sand, may have no clearly defined B horizon; the A horizon passing directly into an unweathered C horizon.

It is clear, therefore, that the physical and chemical properties of soil can differ greatly, not only from place to place but also from ground level to the C horizon below. In nature the evolution of the soil is closely related to that of the vegetation it supports and both tend towards an equilibrium or climax, depending on climate and physiographic factors. Agriculture and forestry disturb this equilibrium and may greatly affect the physical and chemical properties of soil, particularly in the A horizon, so that only the more permanent properties such as texture and sub-soil characters remain little affected.

1.3 Soil types

On a world scale, soil can be divided into a number of main types associated with the major zones of climate and vegetation. For example, forests are characteristic of humid regions, i.e. regions where rainfall exceeds evaporation from the soil, and grasslands are characteristic of more arid regions where potential evaporation from soil greatly exceeds rainfall. In humid regions, percolating water leaches soluble constituents, particularly calcium, from the top-soil, which consequently tends to become acid and contain less plant nutrients than deeper soil layers which can be reached by tree roots. In contrast, where such leaching does not occur, as in the grassland soils of arid regions, plant nutrients tend to be more plentiful in the surface layers of the soil where the main root zone of herbage plants occurs.

Coniferous forest is the characteristic vegetation of the cold temperate region and the soil type associated with it is the podzol (Fig. 1–1). In a typical podzol there is a surface O horizon, consisting mainly of partially decomposed leaf litter and other plant remains, and only a thin A horizon overlying a bleached E or A_2 horizon, from which plant nutrients, iron and aluminium compounds have been leached by humic acids percolating down from the organic layer where they are formed. The leached materials are deposited in the B horizon which is usually quite compact—brown, red or black in colour and sometimes very hard. Podzols are typically very acid and may develop in moorland and heathland where the parent rock is acidic, as well as under coniferous forest.

In the deciduous forest zone, the characteristic soil is the brown forest soil. Its A horizon consists of an intimate mixture of well decomposed organic material and mineral soil, usually black or dark brown, which gradually merges into the brown B horizon underneath. The mixture of organic and mineral material, often called 'mull', after the Danish soil scientist who first described it, is considered to result from the action of earthworms which are plentiful in this soil type, and bury and eat plant remains and soil, producing an intimate mixture of humus and soil in the form of wormcasts. This contrasts with the type of humus formation in podzols, where the organic remains are less decomposed and form a separate layer; earthworms being scarce in such acid soils.

The soil type characteristic of steppe vegetation in semi-arid regions is the chernozem; the soil profile consists of humified organic matter intimately mixed with the mineral soil to give a blackish A horizon that may be one or two feet thick. Much of the organic matter in chernozem soils is derived from plant roots rather than aerial plant remains, and grass roots play a dominant part in the formation of chernozems.

In Britain, the climate tends to aid in making organic matter accumulate as peat in cold, humid, upland areas where the soil is strongly leached and plant remains decompose slowly. However, the major differences in soil are usually related to differences in the parent material and in relief. For example, soils derived from chalk tend to be light, shallow, free-draining, rather dry and have a rich fauna; whereas soils derived from Gault, Oxford clay and chalky boulder clay tend to be heavy, wet and liable to water-logging; and soils derived from coarse sandstone, such as the Lower Greensand in southern England and the Carboniferous limestone in the Pennines, tend to be very acid and have podzolic profiles where they carry heath or coniferous woodland. Soils derived from glacial drift reflect the mixed origin of the parent material and may be quite unrelated to the rocks beneath them.

Other differences can result from relief or landscape history. For example, poorly drained (gley) soils tend to occur in valleys and low lying sites whatever the parent material, and peaty soils may occur where the water table has been at or near the soil surface for significant periods in the recent past.

1.4 Soil pore space, texture and structure

Soil is porous; the solid particles usually occupy 40–70% of the total soil volume, depending on the horizon and the soil type. The pore space between the solid particles is filled with water or air and provides the living space for the soil organisms. In sandy soils pore space may be 30% of the total soil volume, in clay soils it may exceed 50%. In some old grassland soils it exceeds 50% but is less in arable land.

The amount and kind of pore space depends on the soil texture, which is determined by the relative proportions of the different-sized particles present, and on the soil structure, which is determined by the way the individual particles are aggregated. Table 1 shows the fractions into which the mineral particles are conventionally divided according to their equivalent diameter (none is precisely spherical) and the proportions of those fractions in soils of different texture.

Table 1. Particle size and texture of soil.

Fraction	*Stones*	*Coarse sand*	*Fine sand*	*Silt*	*Clay*
Equivalent diameter (mm)	> 2	2–0·2	0·2–0·05	0·05–0·002	< 0·002
Soil texture			Sand	Silt	Clay
Sandy			> 80%		
Silty				> 50%	< 40%
Loamy			< 80%	< 50%	< 40%
Clayey					> 40%

Although the sand and silt fractions have important effects on air and water movement in soil, they are much less active chemically than the clay fraction. Because of the enormously greater surface area of the clay particles per unit mass, much of the chemical and biological activity of the soil depends on the clay fraction. The clay particles also contain minerals that are important sources of plant nutrients—such as calcium, sodium, potassium, magnesium and iron—and, when in combination with humified organic matter, are important in determining the structure of the soil.

Although soil structure is more difficult to specify than soil texture it is readily recognized as the shape and size of the lumps into which a soil disintegrates when it is crumbled lightly. In some soils these lumps may be plates, prisms or blocks with spaces between them when they are dry and have shrunk, but which fit together closely and leave little intervening space when they are wet and swollen. In others the lumps are granular and produce an open crumb structure that drains well and is well aerated. The precise way in which the lumps are formed is not fully understood. Soil texture plays a part, as do colloidal cementing agents such as iron and aluminium hydroxides. The soil organic matter is also important, especially the humified material that has undergone considerable microbial decomposition, lost all traces of structure, thus occurring as an amorphous complex that is very resistant to further decomposition. The formation of a crumb structure, which is characteristic of grassland soils, is particularly associated with the presence of plant roots and the build up of humified organic matter.

1.5 Soil water

In well-drained soils, the pore space is seldom completely filled with water. If it is waterlogged, anaerobic conditions soon develop because of oxygen lack. If a permeable soil is saturated with water and then allowed to drain freely, water drains by gravitational force from the large pore spaces which fill with air, but is held by capillarity in the smaller pores. Soil in this condition is at 'field capacity', i.e. it holds as much water as it can under free drainage. Water can be withdrawn from the pore spaces by evaporation, plant roots and by animals, until the force they exert fails to overcome the force with which the water is held in the pore capillaries. When this force exceeds about 10 atmospheres, plants cannot extract enough water to maintain turgor and they wilt; when the force exceeds about 30 atmospheres they can no longer extract any water and they die. The soil may, however, still hold a good deal of water in the finest capillaries or adsorbed on soil colloids, depending on the texture and composition of the soil. The amount can be measured by drying the soil. Clearly, the moisture content of soil measured by drying does not indicate the availability of the soil moisture to plants and soil animals. A better indication is given by the suction with which water is held by the soil. This is usually expressed as the soil pF—which is defined as the logarithm of the suction force, expressed in centimetres of water, with which the soil moisture is in equilibrium. Unfortunately, the soil pF is rather difficult to measure.

1.6 Soil air

The volume of air in the soil is controlled by the pore space and the amount of water present. In heavy clay soils with a fine texture which are able to hold much moisture, only a small proportion of the total soil volume may be air. By contrast, in old grassland soils with a good crumb structure, about 20% of the total soil volume may be air, even when the soil is at field capacity. On average there is no gross difference between the composition of the soil air and the atmosphere, because gaseous diffusion is quite rapid. However, there can be considerable local differences in soil because of plant and animal respiration, which locally increases the carbon dioxide concentration and decreases the oxygen concentration. Air in the deeper soil layers usually contains more carbon dioxide than the atmosphere. The difference between the humidity of the air in the soil and in the atmosphere is of great importance to soil organisms. The soil air is usually saturated except in the loose superficial layer of cultivated or exposed soil, or in extreme drought. Consequently the soil is a suitable environment for organisms that need a saturated atmosphere as well as for those that need free water.

1.7 Soil temperature

The temperature at the surface of exposed soil closely follows the variations in air temperature but a cover of vegetation lessens daily and seasonal variations in the soil temperature, which becomes less variable with depth. Below about 8 in. there is almost no diurnal variation in temperature, and seasonal variation is much less than near the soil surface. Soil temperatures fluctuate less in clay soils and wet soils than in sandy or dry soils.

Organisms and their Ecology 2

2.1 Microorganisms

It is convenient to divide soil organisms into two major size categories. Microorganisms can be studied satisfactorily only with a microscope; they include bacteria, actinomycetes, protozoa, fungi and algae. Viruses, although not strictly organisms, can multiply, and are therefore usually included in the study of microorganisms, although most are visible only

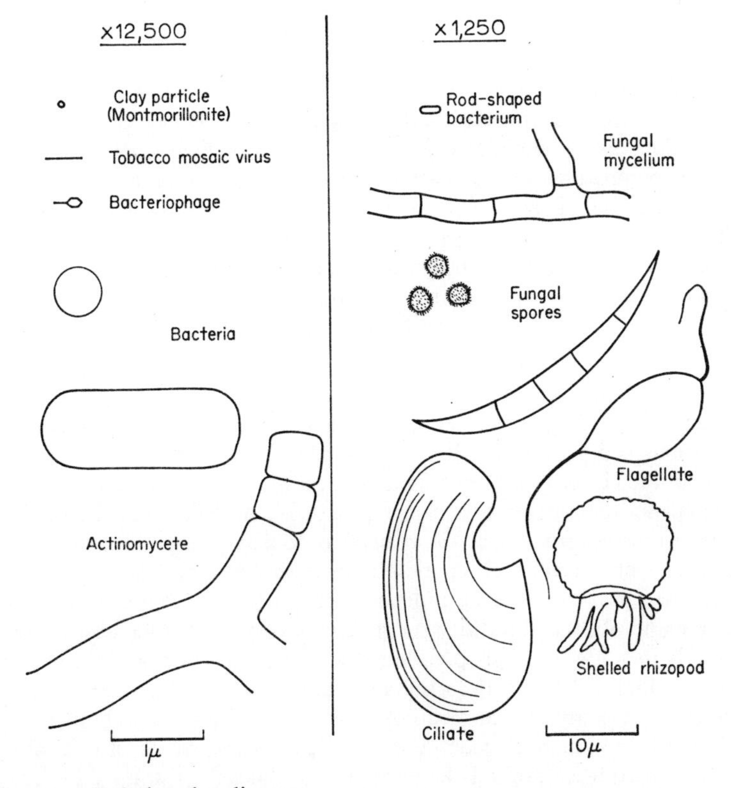

Fig. 2–1 Relative size diagram.

under an electron microscope. The remaining organisms are the soil animals with the exception of the protozoa. Although some, such as nematodes and mites, are small, many can be studied with the naked eye and some are even quite large.

2.1.1 *Viruses and bacteriophages*

Viruses are the smallest and simplest multiplying entities and have an importance in soil because of their effects on complex organisms. All are obligate parasites, and although they may be able to survive outside their hosts, they can multiply only within living cells. Some of the more stable plant viruses, tobacco mosaic virus for example, may remain infective and quiescent in soil for several months or longer. Other plant viruses, that have been described as being soil-borne, are known to be transmitted from one plant to another by soil organisms, such as nematodes and fungi. There is no evidence that these vectors are harmed by the viruses they can carry.

The group of viruses that parasitize bacteria and actinomycetes are called bacteriophages, or simply phages. Some have a more complex structure than plant viruses; these consist of a head with a central core of D.N.A., and a protein coat and tail. When a bacterial cell has become infected, the phage particles multiply within their host, which finally ruptures, releasing large numbers of new phage particles. It is not known what effect phage has on populations of soil bacteria, but whenever a particular bacterial species is present in soil, its phage can usually also be found. Phages possess different degrees of host specificity; some are able to infect several different species of bacteria, others only one species, and some are so specific that they can infect only one strain of a particular species. Phage specificity is used as an aid to the identification of bacterial isolates.

2.1.2 *Bacteria*

Bacteria are the most numerous organisms in soil and of great importance in various soil processes. Many are beneficial to plants, playing essential parts in the circulation of nutrients. Others are harmful, either competing with plants for nutrients or causing diseases. A gram of fertile soil may contain more than 10^9 bacteria, equivalent to a live weight of over 3,000 lb per acre, but infertile soils contain many fewer. Attempts have been made to use bacterial counts as an indication of soil fertility, but the correlation between numbers and fertility is not usually very good. The most common bacteria in soil are rod-shaped, less than 1 micron wide, and up to a few microns long (Plate 1 shows the appearance and relative sizes of typical bacteria, actinomycetes and fungi). Many swim about actively in the soil solution by means of flagella, fine whip-like organs, borne in tufts at one end, or distributed around the periphery of the cell. Also outside the cell wall and secreted by the cell there may be a sheath of polysaccharide gum, called a capsule (Plate 2). The function of this capsule is not certain, but it may protect the bacteria from ingestion by predators such as protozoa. The presence of large numbers of capsule-forming bacteria is thought to have the effect of improving the crumb structure of the soil by cementing together mineral particles and humus.

The ordinary vegetative cells of bacteria are neither particularly resistant to heat nor to desiccation. However, some species of bacteria can form thick-walled spores within their vegetative cells. These endospores can withstand drying for very long periods and are much less susceptible to heat than ordinary vegetative cells; some being able to survive for an hour at 100°C. Spore-forming bacteria of the genus *Bacillus* are very common in most soils. Numbers of bacterial spores present in a soil sample can easily be counted using the method described for counting bacteria (3.3.1), but pasteurizing the initial soil suspension. This is done by immersing the tube or flask containing the initial suspension in a beaker of water held at approximately 80°C for 10 minutes before making the necessary dilutions. Most soil bacteria require oxygen for growth, and are therefore classed as aerobes; if they can also live without oxygen, they are facultative anaerobes. Those that can not live in the presence of oxygen are obligate anaerobes. Many aerobic soil organisms use oxygen so rapidly that anaerobic conditions develop locally even in well-drained soils. In waterlogged soils anaerobic conditions may develop in most of the profile.

When conditions are favourable, many bacteria multiply very rapidly; some divide as often as once every 20 minutes, but most soil bacteria take much longer. Very rapid rates of division cannot be maintained for more than short periods, because the nutrients are soon exhausted, but they do allow bacteria to take immediate advantage of any new food source and so compete successfully with other organisms.

Soil bacteria range widely in their nutritional behaviour, but they divide into two major nutritional groups, the heterotrophs and the autotrophs. Heterotrophic bacteria require organic compounds to supply their needs for carbon and energy, as do fungi and animals, whereas autotrophic bacteria can get their carbon from carbon dioxide and their energy either from oxidizing inorganic substances or from sunlight. Heterotrophic bacteria are numerically the more important, but, as we shall see, autotrophic bacteria perform certain chemical changes of great significance for plant growth. The heterotrophic bacteria include those with the ability to use many different organic compounds, including sugars, cellulose, chitin, organic acids, alcohols and hydrocarbons. It is fortunate that there are bacteria that can decompose many of the organic herbicides and pesticides that are now being used in such large quantities. However, some of these compounds resist decomposition by soil bacteria and may therefore accumulate in soil.

Nitrogen, which is essential for all organisms, can be obtained in different ways by soil bacteria. Some require complex nitrogen compounds but most can use ammonium salts or nitrates. Several species of bacteria are able to make use of atmospheric nitrogen by the process of nitrogen fixation; the importance of these bacteria in the circulation of plant nutrients is discussed in Chapter 4 (4.3.5, 4.3.6).

Examples of autotrophic bacteria are nitrifiers (4.3.2), sulphur oxidizers (4.5) and iron bacteria. The last oxidize ferrous to ferric salts; they do not cause iron, buried in soil, to corrode.

2.1.3 *Actinomycetes*

These are organisms that have some superficial similarities to fungi, but are now generally considered to be highly-evolved and complex bacteria. In neutral or alkaline soils they may be as abundant as bacteria, but they cannot tolerate acid conditions. The vegetative stages of these organisms consist of a fine branching network of filaments, less than 1 micron in diameter with few cross walls. Reproduction is by fragmentation of the filaments into small, rod-shaped pieces, or more commonly, by producing spores, sometimes singly, sometimes in chains on special branches. The spore chains formed by the genus *Streptomyces* are often spirally-coiled (Plate 3). Some actinomycetes form small motile spores. Many can grow at high temperatures and thus play an important part in the fermentation of composts and manures. Many soil-inhabiting actinomycetes produce antibiotics, which inhibit bacteria or fungi, and help them to compete in the soil with other organisms. An example is *Streptomyces griseus* which produces the medically valuable antibiotic streptomycin. Common scab disease of potatoes is caused by another species of *Streptomyces*, but most are saprophytes. In culture, actinomycetes often have a strong and distinctive smell, which is thought to be partly responsible for the characteristic earthy smell of soil.

2.1.4 *Protozoa*

It is only to be expected that soil contains animals that prey on the bacteria living there. Much the most important of these predators are protozoa, many of which feed exclusively on soil bacteria. Three classes of protozoa are well represented in soil: rhizopods, flagellates and ciliates. The rhizopods or amoebae are of two distinct kinds, those that consist of a mass of protoplasm surrounded only by a flexible pellicle; and those with a rigid shell, composed of chitin or silica, from the open end of which emerge fine protoplasmic strands. Both kinds of amoebae can form resistant chitinous cysts when the environment becomes unfavourable for their growth.

The flagellates are organisms that swim about in the soil solution by means of one or more flagella attached to one end of their bodies. Ciliates are also free-swimming, but differ from flagellates in that they have many cilia, often arranged in bands. These propel the animals and may also assist in feeding by sweeping food into the mouth or cytosome.

Counts of protozoa (3.3.7) have usually shown that amoebae and flagellates are much more abundant in soil than ciliates. For example, one gram of a sample of an English wheatfield soil was found to contain 1,500 amoebae, 32,000 flagellates and only 20 ciliates. It is possible that the relative abundance of the different classes obtained in these counts may have been due partly to the particular method used. When different methods were used to study the protozoa in several New Zealand soils it was found that ciliates were always well represented, even in dry soils.

Most soil protozoa encyst when they are short of food or conditions are otherwise unfavourable. Cysts are much more resistant to heat and drying than the active forms and they are easily dispersed in dust by wind over considerable distances. Active protozoa have been recovered from cysts that have survived in soil that has been kept dry for 49 years. As cysts resist acid, it is possible to make differential counts of cysts and active forms by estimating numbers (3.3.7) before and after treating the soil for a few hours with 2·0% hydrochloric acid.

Bacteria form the food of most soil protozoa, but fungi are also eaten (Plate 4). Not all bacteria are equally suitable; some, including strongly pigmented bacteria such as *Serratia* are inedible and a few are toxic to protozoa. In addition to the predatory protozoa, some are saprozoic and feed on dead organic matter. Some flagellates contain chlorophyll and can photosynthesize; these are most conveniently regarded as algae, but except for their photosynthetic pigments are identical with other flagellates.

2.1.5 *Fungi*

Fungi are probably of equal importance to bacteria in contributing to soil processes and plant nutrition in neutral and alkaline soils. Fungi usually tolerate acid conditions better than bacteria, and for this reason are more important than bacteria in acid soils. Estimation of the amount of fungi in soils has special problems not set by bacteria, and some of these are discussed in the section on methods (3.3.5). Plate counts of fungi in normal soils give figures of up to several hundred thousand per gram, giving the impression that they are less abundant than bacteria (2.1.2). A better comparison can be made between the estimated volume of bacterial and fungal tissue in soil. The volume of fungal tissue can be calculated from the length and average diameter of hyphae in a gram of soil. Average figures for pasture soils have been found to be about 100 metres in length with an average diameter of 5 microns. This gives a volume of approximately 2,000 million cubic microns per gram of soil, or about 0·2% of the soil volume. A similar calculation for soil bacteria, assuming that each bacterium has a volume of one cubic micron, gives a figure close to that for fungi. Of course a comparison of this sort still tells us nothing about the relative biochemical activities of the two kinds of organisms.

Soil fungi are of many different types, ranging from primitive unicellular kinds to the toadstools with their large and complex fruiting bodies. The class of fungi known as the Phycomycetes includes several genera of saphrophytic moulds common in soil, for example *Mucor*, *Absidia* and *Mortierella*, but the class also includes plant parasites like *Phytophthora infestans* (the cause of potato blight) that live in the soil for at least part of their life cycles. Of somewhat uncertain systematic position, but usually included in the Phycomycetes, are fungi belonging to the genus *Endogone*. Some members of this genus are saprophytes, but others that are very abundant in soil, are obligate symbionts of higher plants. When the spores of these fungi germinate, they can grow only to a very limited extent unless their germ tube penetrates a plant root. After penetration, an extensive mycelium develops outside the root, and in the root hyphae grow between the cells of the cortex and produce side branches that enter the cortical cells (Plate 5). Inside these cells the branches form either vesicles or much-branched structures called arbuscules. The association of fungus and root seems to be a true symbiosis. The fungus depends on the root for further growth, and the plant usually benefits, probably as a result of an increased supply of mineral salts (4.4), conducted from the soil into the roots through the fungal hyphae. The fungus sometimes becomes parasitic and then harms rather than benefits the plant. This close association of fungus and root is a form of mycorrhiza, which, because the fungus develops inside the root and does not form a thick mantle on the outside of the root is called endophytic. Endophytic mycorrhizas of the *Endogone* type occur in a wide range of plants, and with a few possible exceptions there does not appear to be any specialization. Thus, a strain of *Endogone* isolated from strawberries, infected clover, grass, onion and hemp, and could probably have infected many other plants.

The second class of fungi, the Ascomycetes, contains several species common in soil. Members of the genus *Chaetomium* are examples that frequently attack cellulose-rich material in soil. The sexually produced spores of many Ascomycetes are much more resistant to heat than vegetative hyphae or asexual spores and may even require heat treatment before they will germinate. For this reason it is possible to isolate selectively several different kinds of Ascomycetes from soil that has been heated enough to kill the hyphae and asexual spores of other fungi.

Although many Basidiomycetes are common in soil, we know little about their mode of existence there. This is chiefly because they are rarely isolated from soil by ordinary methods; special techniques such as the hyphal isolation technique (3.3.5) are thus needed. We are normally only aware of these fungi when they produce their large and conspicuous fruiting bodies above ground. These structures are really the climax to a long period of activity in soil, often by a very extensive mycelium. There is more knowledge about some species than others, either because they are parasitic on plants, for example *Armillaria mellea*, or because they form ectotrophic mycorrhiza in association with the roots of certain plants. This type of mycorrhiza differs considerably from that formed by *Endogone*. Infected roots are typically short and frequently branched. They are

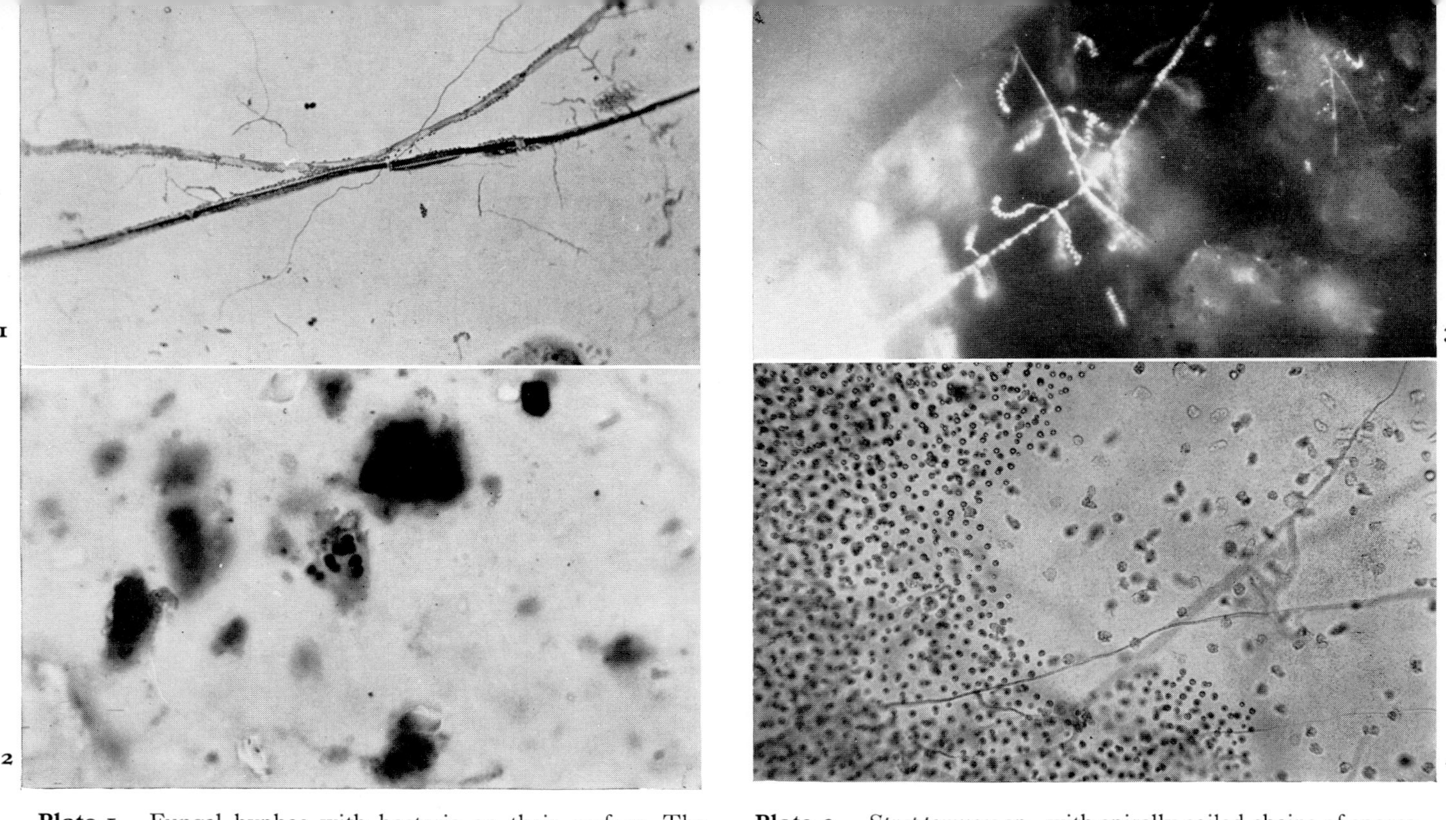

Plate 1. Fungal hyphae with bacteria on their surface. The fine hyphae are actinomycetes. (× 250)

Plate 2. Micro-colony of large bacteria with paired cells embedded in polysaccharide in agar film preparation. (× 900)

Plate 3. *Streptomyces* sp., with spirally coiled chains of spores, growing on soil particles. (× 175)

Plate 4. Soil amoebae (right) feeding on a yeast, *Lipomyces* sp. (left). No yeast cells remain in the area where the amoebae have fed. Fungal hyphae can be seen crossing the field. (× 125)

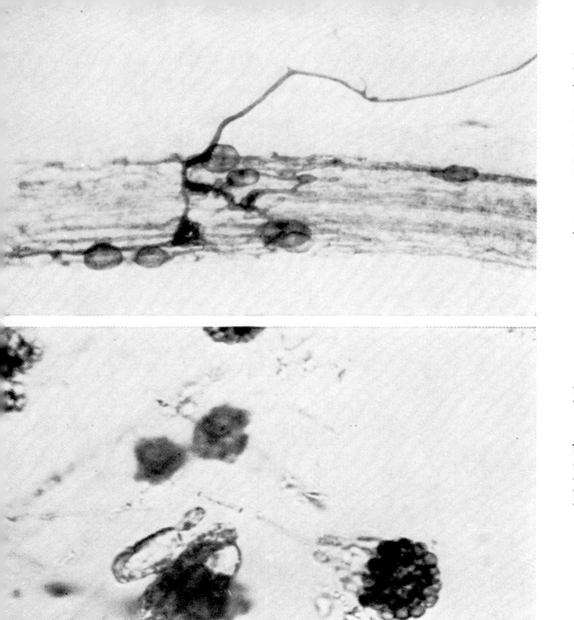

Plate 5. Clover root infected by vesicular-arbuscular mycorrhizal fungus. A coarse hypha has entered the root and formed vesicles and inter-cellular hyphae. (× 100)

(By permission of Dr. B. Mosse, 1963. *Symbiotic Associations*, Cambridge University Press.)

Plate 6. *Gliocladium* sp. (Fungi Imperfecti) sporulating on soil particles. Spores form a slime-enveloped head at the apex of branched fertile hyphae. (× 375)

Plate 7. Radula track of slug. (From H. F. Barnes, 1950, *Bedfordshire Naturalist*, 5.)

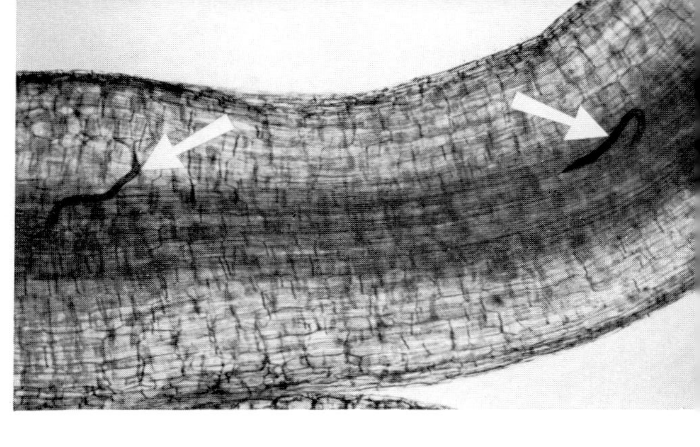

Plate 8. Potato cyst nematode: (a, above) infective larvae soon after entering root; (b, below) cysts on potato roots. (By courtesy of C. C. Doncaster.)

covered by a thick mantle of fungal hyphae from which some of the hyphae penetrate the cortex of the root and others grow into the soil. Ectotrophic mycorrhizas are common on trees; different species of trees form them only with certain species of Basidiomycetes. For example, birch trees form mycorrhiza with *Boletus scaber* and the fly agaric, *Amanita muscaria*; consequently fruiting bodies of these fungi are frequent under birch trees. In fertile soils ectotrophic mycorrhiza seem to have little effect on tree growth, but in infertile soils trees with mycorrhiza are usually more vigorous than those without.

Many of the common fungi isolated from soil produce abundant asexual (imperfect) spores (Plate 6), but have no known sexual stage, or do not produce it easily in culture. Most of these fungi are the asexual stages of Ascomycetes, but they have to be classified by their imperfect sporing stages, and are therefore placed in an artificial group called the Fungi Imperfecti. Species of the genus *Penicillium*, many of which produce antibiotics, are common examples of this group in soil.

The yeasts are a group often studied apart from other fungi, although most are either Ascomycetes or Fungi Imperfecti that have in common the characters of not forming a true mycelium and of reproducing asexually by budding. Soils generally possess a distinct yeast flora, consisting of species not common in other environments. Sometimes a single species may become very abundant. For example, more than a million cells of the imperfect yeast *Candida curvata* may occur in each gram of soil. Leaf surfaces also support characteristic species of yeasts, and these may be washed by rain into the upper layers of the soil.

Soil fungi can be either saprophytes, parasites or symbionts. Like soil bacteria, saprophytic soil fungi can use as food a very wide range of organic compounds. A major constituent of wood and all secondarily thickened plant tissues is lignin, which is decomposed very actively by many fungi, though by few if any bacteria. For this reason fungi are the primary organisms responsible for the breakdown of all woody tissues. Many kinds of fungi decompose cellulose and are often the first organisms to attack this substance when it is added to soil. Keratin and chitin can also be decomposed by several kinds of fungi.

Most soil fungi contribute to the essential processes of decomposing organic matter in soil, but others are harmful, causing economically important root diseases. Young seedlings, which are particularly susceptible to fungus attack, can often be protected by treating the seed before sowing with a fungicidal compound that prevents fungal invasion of the young root. Infection of the roots of older plants is more difficult to prevent, except in glasshouses where it may be practicable to treat the soil either with chemicals or steam, before planting, to kill any pathogenic fungi present.

In addition to fungi that parasitize plants, there are some that attack and parasitize soil animals, particularly nematodes. These predacious fungi have evolved various mechanisms for catching their prey. Some form hyphal nooses that contract and strangle nematodes, others have sticky pads that catch nematodes. These fungi are thought to help in controlling natural populations of plant-parasitic nematodes, but attempts to control nematodes by inoculating soil with predacious fungi have not been very successful.

2.1.6 *Algae*

Algae differ from the other organisms so far discussed by having photosynthetic pigments that enable them to synthesize carbon compounds from carbon dioxide in the presence of light. The need for light gives an immediate clue to their distribution in soil; they are indeed found most abundantly either on the soil surface, if conditions are sufficiently moist, or just below the surface. Also, because of their requirement for light, algae develop most abundantly when the soil is not heavily shaded by vegetation or surface litter. Although never as numerous as near the surface, algae do occur at depths of 20 cm or more. Some of these may have been washed down from the surface through the soil, particularly if it has an open sandy texture. Others, however, can grow heterotrophically in the dark and may therefore be true inhabitants of the deeper soil. Heterotrophic growth of algae, dependent on organic carbon compounds, is rarely as vigorous as autotrophic growth.

Green algae (Chlorophyceae) are the group most commonly represented in soil. Genera often found include *Hormidium*, *Chlorella* and *Chlamydomonas*. The blue-green algae (Cyanophyceae) are also common and under some conditions form continuous mats on the soil surface. Most luxuriant development is associated with very damp tropical conditions. This group has special significance because many of its members are active nitrogen fixers (4.3.6). The characteristic empty silicaceous shells of diatoms (Bacillariophyceae) are often seen during microscopic examination of the soil, and indicate that live diatoms must have been present. Yellow-green algae (Xanthophyceae) are less common in soil than the other groups mentioned and red algae (Rhodophyceae) are rarely found except in abnormal saline oils.

2.2 Macrofauna

The soil contains a remarkably diverse population of animal life, ranging from the smallest organisms such as protozoa to large burrowing vertebrates such as rabbits, badgers, marmots and many others. Some animals, such as earthworms, symphyla, protura, some mites and collembola, spend the whole of their life cycle in the soil and are the permanent soil fauna. Others, such as burrowing rodents, reptiles and amphibia and many insects, spend only part of their time or only specific stages of the life cycle in the soil, and are the temporary soil fauna.

Soil animals may be grouped in other ways. For example, burrowing animals such as earthworms, millipedes, rodents, ants, termites, fossorial beetles, which can create their own living space by burrowing in the soil; mites, collembola and many insect larvae, cannot burrow except possibly to a very limited extent so they live in existing air spaces in the soil; rotifers, ciliates and protozoa need free water or a water film for movement so they are confined to the water film surrounding soil particles. Some animals such as slugs, earthworms, enchytraeids and some insect larvae require moist conditions and so are intermediate between the water fauna and the air fauna.

Again, the soil fauna can be classified according to feeding habits. Phytophagous forms, which feed on living plant material, include many important crop pests such as slugs and snails, symphyla and many insect larvae and adults, e.g. wireworms. Saprophagous animals, which feed on dead plant material, include many such as earthworms, millipedes and many insects such as ants, termites and some fly larvae that are important in the decomposition of soil organic matter. Carnivorous animals include moles, centipedes and many beetles and their larvae, e.g. carabids and staphylinids. Several other sub-divisions could be made. A census of the arthropod population gives some idea of the structure and spatial distribution of the soil population. Although the arthropods are only part of the total soil population they do illustrate some general features of it.

Figure 2–2a summarizes the results of a census in a permanent pasture used for grazing. Twenty soil samples, each 4 in. diameter and 12 in. deep were collected in November and the arthropods extracted using the flotation method described in section 3.4.2. Each sample was divided so that the arthropods in the top 6 in. were extracted separately from those in the lower 6 in. The collection was known to be incomplete because the finest sieve used had a mesh of 0·1 mm across and could not retain the smallest arthropods such as larval mites and collembola.

Even so, enormous numbers of arthropods are shown to occur in the soil Most of them of course are very small and together would probably not occupy more than 1 part in 5,000 of the pore space in the top-soil of a pasture.

The results illustrate the diversity of classes and orders of arthropods found in soil, and when they are arranged in approximate order of size (Fig. 2–2a) there is a pyramid of numbers with many smaller animals and fewer large ones. Similar pyramids could be constructed for other taxonomic groups, e.g. the Oligochaeta, and a similar pyramid undoubtedly exists for the total soil fauna and flora, with viruses or bacteria at the base and burrowing vertebrates at the top.

Of the arthropods 70% were in the top 6 in. However, the vertical distribution is not the same for all groups. No spiders and few lepidoptera larvae were found below 6 in. but symphyla, protura and pauropoda, all part of the 'permanent' soil fauna, were more abundant below 6 in. than

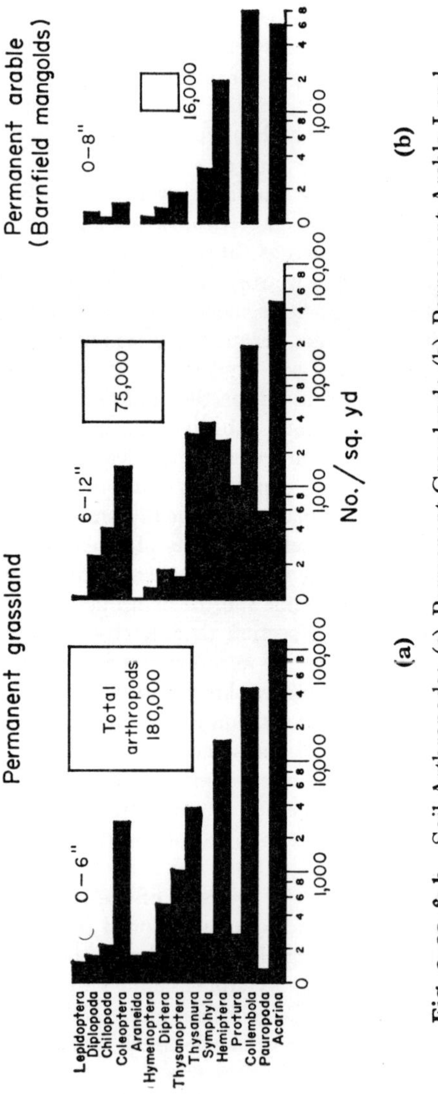

Fig. 2–2a & b Soil Arthropods: (a) Permanent Grassland; (b) Permanent Arable Land.

above. Smaller organisms tend to be relatively more abundant deeper in the soil. Such vertical stratification of the population is characteristic of the soil fauna, and probably reflects the distribution of food supplies (soil organic matter) and possibly changes in pore space and soil atmosphere.

A similar census (Fig. 2–2b) was made by the same method but in an old arable field that had grown a root crop each year for more than 80 years. Some groups found in the pasture soil, such as lepidoptera larvae, spiders, thysanura and protura, were not found in the arable soil: collembola, not mites, were the most abundant group and the total number of arthropods was one twentieth the number found in the pasture. This shows that the diversity and abundance of the soil fauna can change with cultivation. Similar studies of the soil fauna, or certain groups in it, could be made in other habitats using some of the methods described in Chapter 3.

2.2.1 *Mollusca*

Although slugs and snails are common, many aspects of their biology and ecology are imperfectly known and would repay further study. Most species can be identified from external characters; snails can often be identified by their shells, and good keys are available. As there are only about 25 British species of slugs it is quite possible to know all of them. Unfortunately, there is no convenient and accurate method for estimating their abundance, and those commonly used measure activity as well as abundance. One way is to count the number or collect those seen on, say a 30-minute walk, after dark, following a predetermined path through the habitat, e.g. a garden. Another way is to trap specimens with metaldehyde and bran pellets. Either method could be used to show the species in different habitats, to study activity in relation to weather, or the seasonal occurrence of particular species and stages.

Slugs and snails can be reared easily (3.5) and used to study reproductive rate, incubation period of the eggs, or growth rate and longevity. Many species are omnivorous and some such as *Testacella* spp. are carnivorous so there are many possibilities for studying feeding habits and food preferences, and the effect of different foods on growth rate and reproduction. For quantitative studies, foods can be supplied as leaf discs or discs of carrot, potato, turnip, etc., cut to uniform size, with a cork borer and sliced to uniform thickness. The amount eaten can be measured, or results can be converted to dry weights (see also 2.2.4.).

Slugs and snails feed in a characteristic way by rasping their food with the radula, a toothed chitinous structure situated in the buccal mass. This leaves a characteristic pattern that can be recorded for different species in the following way.

Coat the inside of several 3½ in. petri dishes with a thin layer of hard beef or mutton fat by heating a small quantity of fat in them and then letting them cool. Put a few specimens of different slugs and snails into

1 lb jam jars or honey jars containing a few drops of water, using a different jar for each species, and then cover each jar with one of the Petri dishes. Examine the Petri dish each day for the imprint of the radula teeth in the fat and see if the pattern for each species is distinctive (Plate 7).

The radula of each species can then be extracted and mounted in the following way. Remove the head of the specimen and let it rot in a dish of water. Pick out the radula, rinse it well to remove remains of tissue and spread it out, teeth uppermost, in a drop of water on a microscope slide. Place a cover slip over it and flatten the radula by weighting the cover-slip if necessary. Allow the mount to dry and then carefully remove the cover-slip. The radula will remain attached to the slide, where it can be mounted in glycerine jelly and ringed. With preserved specimens the buccal mass should be dissected out and left overnight in 10% caustic soda solution. The radula can then be removed, rinsed and mounted as above.

Much of the food of slugs and snails is fresh or decomposing plant material containing celluloses and hemicelluloses that many soil animals, such as the arthropods, can use only if their gut flora contains micro-organisms, such as bacteria and protozoa that secrete cellulases. However, the digestive juice of molluscs contains powerful cellulases that the animal itself secretes. Cellulose digestion may be demonstrated as follows:

Collect several mature specimens of *Helix aspersa* or *H. pomatia*, starve them for several days and then kill them with chloroform and remove the shells. Open up the snails with a slit on the left side near the edge of the mantle so that the crop protrudes or can be pulled out with forceps. Then cut the oesophagus and intestine and collect the drops into a small beaker. Cut up the crops roughly with scissors and add a little water so that there is about 1 ml of extract for each snail. The extract should be clarified by centrifuging or by allowing it to stand overnight in a refrigerator and then filtering the supernatant fluid through a coarse filter paper.

Cellophane tubing, as used for dialysis, is a suitable form of cellulose substrate. Cut six equal lengths weighing approximately 100 mg and weigh each accurately. Soak them in water overnight to remove soluble plasticisers. Reweigh two of the pieces after drying at 100°C and use the results to calculate the dry weight of the other four pieces. Split these open and crumble two into test tubes containing 1 ml of the snail digestive juice extract, 1 ml of water and 2 ml of acetate buffer (pH 5, made by mixing 1 part N/10 acetic acid with 4 parts N/10 sodium acetate). Put the other two pieces into tubes containing buffer solution only and set up two further tubes containing digestive extract and buffer solution only. Incubate the six tubes for 24 hours at 30–35°C. Remove the undigested pieces of cellophane, wash them and weigh them after drying at 100°C. The loss in weight during incubation gives the amount of cellulose digested. The presence of soluble carbohydrate can be tested qualitatively in each of the tubes or estimated quantitatively by one of the standard methods such as the iodine–thiosulphate titration method for reducing sugars.

2.2.2 *Oligochaeta*

EARTHWORMS. Gilbert White expressed a now commonly held view of the importance of earthworms when he wrote, 'Worms seem to be great promoters of vegetation which would proceed but lamely without them, by boring, perforating and loosening the soil, and rendering it pervious to rains and the fibres of plants, by drawing stalks of leaves and twigs into it; and most of all, by throwing up such infinite numbers of lumps of earth called wormcasts, which, being their excrement, is a fine manure for grain and grass. . . . '

Later, Muller, studying Danish forest soils drew attention to the importance of earthworms in soil formation. He considered that in 'mull' soils (1–3) the form and distribution of the organic matter was the result of earthworms burying and mixing the organic matter with the mineral soil, especially in their intestines, and in producing water-stable aggregates in the form of wormcasts. In 'mor' soils, on the other hand, where earthworms are scarce, plant remains are much less comminuted and decomposed, and they accumulate as a discrete layer on the soil surface.

Darwin studied two particular aspects of earthworm activity. He observed the way worms pulled different kinds of leaves into their burrows, for food or to line the burrows or cover the entrance. He also measured the amount of wormcasts produced in various fields and localities and estimated the rate at which a stone-free layer was produced and ancient buildings, pavements, chalk and cinders were buried by wormcasts. Surprisingly, he did not point out that the species that throw up wormcasts were not the same as those that bury leaves.

Muller and Darwin, however, studied earthworms in undisturbed habitats such as forests and permanent grassland, not in land under arable cropping. Despite the beneficial effects attributed to earthworms, almost all attempts to demonstrate them experimentally have failed. There may be two important reasons for this. First, many earthworms are killed by cultivation and there are relatively few in arable land compared with permanent grassland or woodland. Secondly, cultivation itself replaces many earthworm activities such as burying and mixing plant remains in soil and improving drainage and aeration. It is significant that almost the only example of increased crop yield on a field scale by introducing earthworms was in some New Zealand pastures, i.e. in uncultivated soil lacking earthworms for geographical reasons though soil conditions favoured them.

There is much recent work on the effects of earthworms on the decomposition of soil organic matter and on the availability to plants of nutrients, especially nitrogen, in soil. The importance of earthworms in removing and fragmenting leaf discs buried in leaf litter or soil has been demonstrated, and how they affect the availability of nitrogen has been studied. When young growing specimens of *Allolobophora caliginosa* were fed on a finely ground mixture of soil and clover litter to give a medium containing 0·2% N, about 6% of the non-available nitrogen ingested by the worms was excreted in forms available to plants. Also, the presence of earthworms in well aerated cultures of moist soil containing leaf litter or dung increased the rate at which oxygen was consumed and ammonia and nitrate accumulated. The increased oxygen consumption was more than could be accounted for by the worms themselves, indicating that the presence and activity of the worms had stimulated the activity of other decomposers such as microorganisms.

Estimates such as these have sometimes been used to assess the importance of earthworms in the cycle of plant nutrients such as nitrogen, but the calculations involve many assumptions that may be inaccurate and cause very great errors. Moreover, earthworms are not alone in making nitrogen available to plants and it is not known how their contribution compares, for example, with that of soil bacteria.

There are about 25 species of earthworm in Britain, and of these about 10 are common in ordinary agricultural land and gardens. The others occur in rather specialized habitats, such as acid hill pastures, semi-aquatic habitats and compost heaps. There is a good key for identifying adults, and it also describes the cocoons. Unfortunately, immature specimens, which usually form most of the population, cannot always be identified with certainty. However, once the species in a population have been determined from the adults, the immature specimens can often be assigned to their species with reasonable certainty by using morphological characters unrelated to sexual maturity. As with slugs, it is quite possible to become familiar with all the British species. However, the distribution of the species is still imperfectly known and much valuable information about their occurrence and relative abundance in particular habitats could be assembled by simple sampling methods described in Chapter 3.

The life history of many species is still inadequately known and laboratory cultures, which are easy to maintain, could provide much basic information on cocoon production, incubation period, growth rate, feeding habits and effects of food supply, temperature and moisture. For example, the growth period from hatching to sexual maturity seems to range from about six months for *L. castaneus* to over a year for *A. longa*, and there are very great differences in the rate different species reproduce. So far as is known, unmated individuals of *Allolobophora* species produce no cocoons, unmated *Lumbricus* species produce non-viable cocoons, but unmated individuals of some other species such as *Bimastus*, *Dendrobaena*, *Eisenia* and *Octolasium* produce viable cocoons, but fewer than mated individuals.

Food supply can greatly affect the reproductive rate and its effect can be studied readily in laboratory experiments, particularly with the smaller species such as *A. chlorotica* and *L. castaneus* that reproduce rapidly.

The following method has proved successful. Set up 12 cultures of each species in 1 lb kilner jars putting five mature worms in each culture. Supply four cultures of each species with farmyard manure, four with fresh cow or sheep dung and four with chopped straw. At the end of three months recover the cocoons from each culture by the flotation method (3.4.3). Results from one such experiment are—

Food	Average no. of cocoons per culture. *A. chlorotica*	*L. castaneus*
Farmyard manure	0·2	8·6
Sheep dung	14·0	76·0
Chopped straw	1·4	12·0

Several earthworm activities such as burrowing, casting, soil mixing and burial of organic matter can be studied or demonstrated simply in the laboratory using cages (Fig. 2–3 see also 3.2.1) made by slotting two sheets of glass into a wooden frame so that they are about $\frac{1}{3}$ in. apart, i.e. far enough apart for the worms to be able to burrow when the cage is filled with soil, but close enough for the burrows to be seen through the glass. The glass plates need to be covered so that the worms are in darkness except when being observed. To observe how different species mix the soil, the cages should be filled with alternating layers of soil of contrasting colour or texture. To observe how they feed on organic matter and mix it with the soil, the cages should be filled with soil, preferably light coloured, and a layer of organic matter such as peat, chopped straw, leaf mould or farmyard manure put on top.

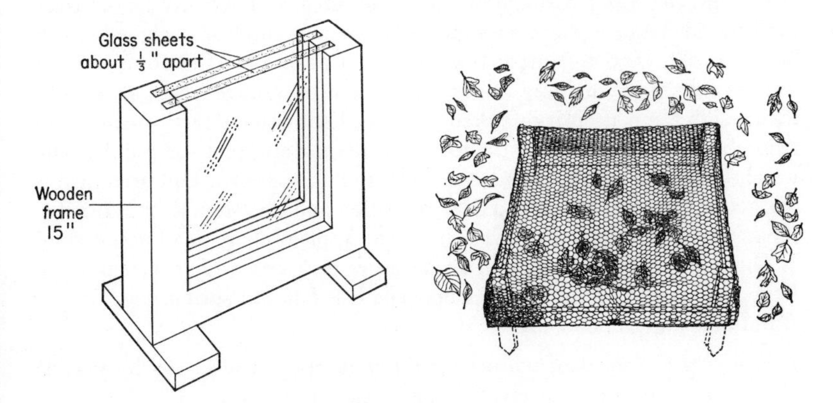

Fig. 2–3 (*left*) Cage for studying earthworm activity.
Fig. 2–4 (*right*) Cage for studying leaf burial by *Lumbricus terrestris*.

The production of wormcasts and the burial of leaf litter can also be studied in the field.

WORMCASTS. Production of wormcasts seems to depend on the number of *A. longa* and *A. nocturna* present. Other species, except possibly *A. caliginosa*, usually cast underground. Wormcasts appear in early autumn and are produced until the worms become dormant in early summer. To study seasonal production of wormcasts, quadrats about one or two yards square should be marked out in the study area and the wormcasts collected regularly and weighed after being air dried or oven dried. At the end of the experiment the earthworm population beneath each quadrat can be estimated by the formalin method (3.4.3) and related to the weight of wormcasts produced. Additional experiments that can be made are—

1. Determine the size of the largest soil particles in wormcasts by crumbling some casts on to a set of graduated sieves and sieving them under water to separate the various size fractions.
2. Mark out quadrats in early autumn or late spring, i.e. before casting starts or just before it finishes. Water half of the quadrats regularly and estimate its effect on the time that casting starts or finishes and the amount of casts produced.

BURIAL OF LEAF LITTER. Of the common species, only *L. terrestris* appears to bury leaves by pulling them into its burrows. This can be observed under trees after leaf fall. It can be studied quantitatively in the following way:—Construct 10 cages about 2 ft square as shown in Fig. 2–4. Select an area where *L. terrestris* occurs, e.g. under trees where leaves are being buried. Collect leaves just after leaf fall, count out 14 batches of 100 leaves, weigh each batch and then oven dry four batches at 80°C and use the results to calculate the oven dry weight of the other 10 batches. Distribute the cages throughout the study area, remove any leaves from the ground beneath them and then scattter one batch of weighed leaves beneath each cage and press the cages into position. After four or eight weeks count the remaining leaves, wash them to remove traces of soil and then weigh them after drying them at 80°C. Estimate the earthworm population beneath each cage by the formalin method and plot the number and weight of leaves buried against the weight of *L. terrestris*. Unburied leaves lose weight from leaching and decomposition and this should be estimated separately. This method showed that some apple orchards had populations of about 1 ton fresh weight of *L. terrestris* per acre, which buried about ½ ton dry weight of apple leaves between leaf fall and spring.
Additional experiments are—

1. Study preferences of worms for different species of leaves by placing known numbers and weights beneath each cage.
2. Record how the number of leaves buried weekly for several weeks is related to soil temperature.

3. Place two or four times as many leaves under some cages as under others and measure the effect on the amount buried.

Some of these experiments can be done in large pots in the laboratory using known numbers of *L. terrestris*.

ENCHYTRAEIDS. The Enchytraeidae are a family of small oligochaetes, usually only a few millimetres long, that are widely distributed in terrestrial, littoral and aquatic habitats. They commonly occur in sewage bacteria beds. They are particularly abundant in organic soils in moist temperate regions; populations of 200,000 per square metre have been recorded from heathland in Denmark and northern England and from coniferous woodland in north Wales. They are less numerous in mineral soils and arable land. The smaller enchytraeids may be confused with soil nematodes but can be distinguished by their lack of buccal stylets, 'teeth', or muscular oesophagus adapted for sucking (2.2.3). The larger enchytraeids, which are commonly found in compost or dung heaps, are distinguishable from immature earthworms because their setae are in bundles of 2, 4 or more, rather than single as in earthworms.

Like earthworms, enchytraeids are hermaphrodite, they reproduce by cocoons, and some species are parthenogenetic. Recently, however, some have been observed that reproduce by breaking into fragments that regenerate into complete worms. Enchytraeids feed on algae, fungi, bacteria and soil organic matter in various stages of decomposition, so they may have an important effect on humus formation, particularly as they tend to be most abundant in acid soils where earthworms are rare. They are sensitive to drought but the cocoons may be more resistant and their survival seems largely responsible for the recovery of enchytraeid populations after drought.

2.2.3 *Nematoda*

Most soils contain enormous numbers of nematodes. Estimates of 10–20 millions per square metre are not uncommon and indeed may be underestimates because of inefficient extraction methods. Although many nematodes are adapted to withstand desiccation temporarily, they are inactive unless the soil has free water. They range from free living saprozoic forms to highly specialized parasites of plants and animals, including man.

The structure of the buccal capsule indicates the mode of life. If specimens collected from soil with a Baermann Funnel are examined under the microscope they can be divided into groups on that basis. A narrow mouth and muscular oesophagus, as in *Rhabditis*, is associated with saprophytic and bacterial feeding. Buccal stylets, found in all plant parasitic nematodes, are for piercing animal and plant cells and feeding on their contents. Strongly developed 'teeth', as in mononchids, characterize predacious species and complex hooks and suckers occur in many species that parasitize animals (Plate 8).

Most nematode species are free-living. Their distribution, movements and population fluctuations in various habitats have been studied, but mostly those species that damage crops or parasitize domestic animals and man. The increasing awareness of the importance of nematodes as crop pests in both temperate and tropical regions has greatly advanced the study of parasitic species. Although they are small and therefore more difficult to identify and study than larger soil animals such as earthworms and insects, they are of great scientific interest because they are so widespread and include species whose mode of life ranges from free-living, polyphagous ectoparasites to host-specific endoparasites, with corresponding morphological and physiological adaptations. The most noticeable of these are loss of mobility, particularly by the females, and the development of resting stages that can resist desiccation, be dispersed widely in soil and on infested plants, and that sometimes become active only in response to exudates from the host plant.

Of the ectoparasites, some are migratory, such as *Ziphinema*, *Longidorus* and *Trichodorus* spp. which have long, fine stylets and feed by piercing the epidermal and cortical cells of a wide range of host plants. Others, such as *Paratylenchus* and *Criconema* spp., are more sedentary and sometimes accumulate round the roots of various host plants in numbers great enough to harm them. Recently, it has been shown that some species of *Xiphinema* and *Longidorus* spread plant viruses of the raspberry ringspot and arabis mosaic type, which have a wide host range and severely damage many kinds of crop. Similarly, some species of *Trichodorus* transmit viruses of the tobacco rattle type that also have a wide host range, including many weed species, and cause diseases of potatoes and legumes. The unselective feeding habits of the nematodes and their feeding mechanism makes them well adapted to be vectors of viruses with a wide host plant range and the phenomenon has interesting parallels with virus transmission by aphids.

The stem and bulb nematode, *Ditylenchus dipsaci*, is a migratory endoparasite with a wide host range that includes oats, rye, legumes, onions, bulbs, especially narcissi and tulips, root crops, strawberries and many common weeds. It occurs in biological races that are morphologically indistinguishable but attack different ranges of host plants. Some plants can be host to more than one race, and the nematodes will also invade plants in which they cannot readily reproduce. The adults live in the intercellular spaces of the parenchyma of the host plant but also migrate through the soil from host to host. In favourable hosts the middle lamellae of the parenchyma cells are broken down by pectinases secreted by the nematode, the cells become rounded, separate and create air spaces in which the nematodes live and reproduce. In unfavourable hosts the middle lamellae may not be broken down, the cells die, or necrotic lesions are formed, and the nematodes cannot reproduce. They feed by piercing the cells and sucking the contents. Even in suitable hosts, their feeding eventually makes the plant tissue degenerate and conditions become less favourable for them. They may then return to the soil or remain in an immature stage which can withstand desiccation for several years. Nematodes at this stage sometimes accumulate in great numbers as 'eelworm wool' between the outer scales of heavily infested narcissus bulbs, and become active again when moistened.

Adaptations to parasitism are most highly evolved in the genus *Heterodera*, which is endoparasitic in roots. The females are sedentary, all or most of the eggs are retained inside them and they swell to a spherical or lemon shape according to the species. After the females die the body wall forms a tough leathery cyst, about 1 mm in diameter, which falls from the roots into the soil and in which the eggs can resist desiccation and remain dormant for several years or until stimulated to hatch by root exudates from the host plant (Plate 9).

Mature cysts of the potato root eelworm, *Heterodera rostochiensis*, are spherical except for a small protruberance at the head end and so can be distinguished from the pear-shaped or lemon-shaped cysts formed by most other species of *Heterodera*. Many cysts of *H. rostochiensis* often occur in the soil of fields, allotments or gardens where potatoes have been grown repeatedly. Various features of the biology and life cycle can be studied with very simple methods, such as the following, which is widely used with cyst-forming nematodes:

Collect a supply of brown cysts of *H. rostochiensis*, containing eggs, from infested soil by flotation as described in 3.4.4. Fill a dozen $3\frac{1}{2}$-in. diameter plant pots about one third full with a sterilized 4 : 1 loam and sand mixture; add 50 cysts to each pot and then fill up with the loam/sand mixture. Plant a small piece of potato tuber, with an eye, in each pot, then sink the pots in a bed of moist but not waterlogged gravel, sand or cinders in a cool place.

Exudates from the potato roots will stimulate the eggs in the cysts to hatch. Second-stage larvae invade the roots just behind the root tips, move through the cells and feed near the endodermis near a group of phloem cells. Feeding stimulates the production of giant cells, which are enlarged multinucleate cells (syncytia) with dense granular contents. The larvae moult three times more before becoming adult. Females, which are distinguishable at the third stage, swell and eventually break through the root cortex but remain attached to the roots by the head end. Adult males leave the roots and live briefly in the soil before they fertilize the females. Immature females are whitish but darken as they mature and eventually turn yellow, then brown after they die and the cyst wall hardens. Mature cysts become distributed in the soil and dispersed in soil, plant roots and other infested material.

By examining the plant roots from two pots each fortnight the stages of the life cycle can be traced. To detect nematodes inside the plant roots the following method can be used:

Wash the roots thoroughly to remove debris, then plunge them into boiling 0·1% cotton blue lactophenol for about three minutes and allow them to cool in the stain. After washing away the excess stain, place the roots in lactophenol overnight to differentiate the tissues and then examine under a microscope. The nematodes will be stained blue but plant tissue is little stained except for meristematic tissue. The free-living stage of the adult males may be found in the soil from the pots.

2.2.4 *Arthropoda*

Myriapods and insects are represented in the soil by so many diverse orders, families, genera and species that they can provide examples of almost every aspect of soil biology. Some general features of soil arthropod populations have been discussed already in section 2·2.

The Pauropoda are part of the true subterranean fauna but little is known about them because they are seldom found except by methods that collect the minutest arthropods. The Symphyla are similarly hypogeal but much more abundant then Pauropoda. They feed voraciously on plant material and soil micro-organisms, and some species are pests of horticultural crops, especially in glasshouses.

The Chilopoda (centipedes) are primarily carnivorous but some geophilomorphs occasionally feed on plant tissues. They are predominantly woodland species but are common also in grassland, arable land and moorland. The species found in moorland and heaths are often those common in woodlands on 'mor' soils whereas those found in grassland and arable land are those commonly associated with 'mull' soils. The local distribution and activity of centipedes depends on their body form and moisture relations. Whereas geophilomorph centipedes can burrow well in soil, lithobiomorph centipedes cannot and are restricted to the litter layer or sheltered places on the soil surface. The cuticle of centipedes has no waterproof wax layer, such as insects have, so they cannot tolerate drought for long, but, because they are so active, then can forage temporarily in dry places that they could not inhabit permanently.

The Diplopoda (millipedes), like the Chilopoda, are predominantly woodland species and those in grassland and arable land are probably relict forest species. Millipedes are exclusively vegetarian and feed on plant material in various stages of decomposition. In places where they are abundant, as in some woodland soils, a humus layer consisting largely of millipede faecal pellets may occur, but their influence on humus formation is less than that of earthworms because they do not ingest mineral matter and produce the clay-humus complex characteristic of the mull humus formation that results from earthworm feeding. Nor do they mix and distribute organic matter throughout the soil horizons as earthworms do.

Millipedes are convenient animals to study in the laboratory and their feeding habits, digestion and effect on soil organic matter have been studied more than most soil animals except, possibly, earthworms. The palatability of different leaf species, in differing stages of decomposition, has been studied by offering millipedes leaf discs and estimating the area or the weight eaten. Faecal pellets can then be collected and weighed to estimate the amount of food assimilated, and chemical analyses done on the food and faeces to study the effects of digestion. The palatability of leaves of different species differs considerably and is affected by the degree of decomposition. As might be expected, millipedes fed on fresh litter eat less but assimilate a greater proportion than those fed on litter that has been weathered and partially decomposed. However, some species, such as *Cylindroiulus nitidus* prefer partially decomposed litter and reject fresh litter. In general it seems that many millipedes eat large amounts of leaf litter of little nutritional value, much of which they excrete relatively unchanged chemically but greatly fragmented and so more readily attacked by microorganisms.

The soil contains many insects that are important crop pests and, as with nematodes, studies of such pests have contributed much to the study of soil animals generally. Examples of well-known crop pests from three different orders are wireworms (Coleoptera larvae), leatherjackets (Diptera larvae) and cutworms (Lepidoptera larvae). These pests live in the soil and feed on plant roots and undergound shoots. Others, such as cabbage root fly, *Erioischia brassicae*, and wheat bulb fly, *Hylemyia coarctata*, also lay their eggs in the soil, but their larvae invade the plant and feed inside the stem or shoots, returning to the soil to pupate. Many other soil insects are predacious, for example, carabid and staphylinid adults and larvae.

Predation in the soil is difficult to study and little is known about the effect of insect predators on populations of their prey. The need for such information has been emphasized in recent years because insecticides applied to the soil as sprays, dusts or seed dressings, or in fertilizers, may kill not only soil-inhabiting pests but also their arthropod predators. The effect of predators on a soil-borne pest, and the way insecticides can alter it, is well illustrated by studies of the beetle predators of cabbage root fly.

In field experiments it was observed that cabbages growing on plots previously treated with DDT, aldrin or BHC were damaged more by cabbage root fly than those growing on plots that had never been treated with insecticides. It was also observed that carabid and staphylinid beetles were less numerous on treated plots for several months after the insecticide was applied. Laboratory feeding tests showed that more than 30 species of beetles, mainly carabids and staphylinids, ate cabbage root fly eggs, larvae or pupae, and that some species common on brassica plots, such as *Bembidion lampros* and *Trechus obtusus*, did so voraciously. By decreasing the number of predatory beetles without controlling the pest, insecticide treatments resulted in more cabbage root fly damage.

Methods of Study 3

3.1 Introduction

The study of soil organisms has many problems not found when studying plants and animals above ground. Most organisms that live in soil spend all or most of their lives hidden from view. Also, many are too small to be seen and studied except with a microscope. These features complicate their study, which has therefore necessitated developing special methods.

It is important to distinguish between methods of study that tell one something about the kinds of organisms present in the soil and those that enable one to estimate their numbers. The former give a qualitative picture that is informative as to the ecology of the organisms, and may enable species lists to be made. The latter are quantitative, giving information about relative numbers of different species present, and the changes in these numbers when conditions change.

3.2 Direct methods

Ideally we would like to be able to observe soil animals directly and microbes in their natural habitat without disturbance; but normally we can only see those organisms that live at least part of their lives at the soil surface. Direct methods have as their objective the study of organisms as they live in the soil.

3.2.1 *Soil observation boxes*

The best way to observe behaviour beneath the soil surface is to enclose soil in boxes with glass sides, which are kept darkened except when observations are being made. Soil animals, fungi and plant roots can be observed through the glass side, either with the naked eye, or with a low-power microscope mounted horizontally. At East Malling Research Station in Kent, a 'root laboratory', that is, a large glass-sided tunnel, has been made, which is used not only to observe the growth of tree roots, but also soil organisms associated with these roots. Slowly-growing objects, such as roots and some fungal hyphae, can be photographed with a time-lapse camera that automatically takes a photograph at fixed intervals of time, perhaps once every minute or every five minutes. When a film taken in this way is projected at normal speeds, movements are much speeded up and show features of growth difficult to observe in any other way; for instance, such films show that root tips wave from side to side 'exploring' their path as they grow. It might be objected that the glass surface in contact with the soil may result in unnatural growth or behaviour of soil organisms, but the observations that can be made this way are often interesting and informative.

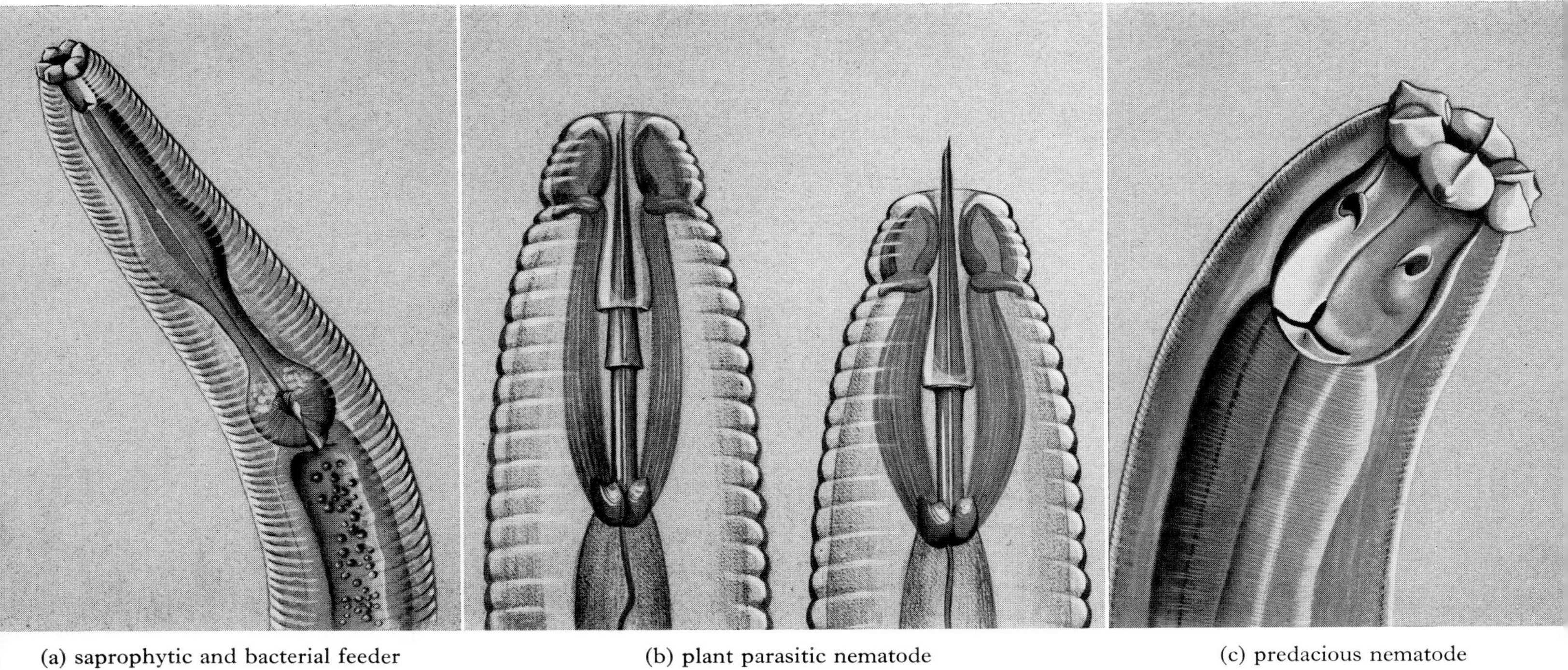

(a) saprophytic and bacterial feeder (b) plant parasitic nematode (c) predacious nematode

Plate 9. Head and buccal capsule of nematodes. (By courtesy of C. C. Doncaster.)

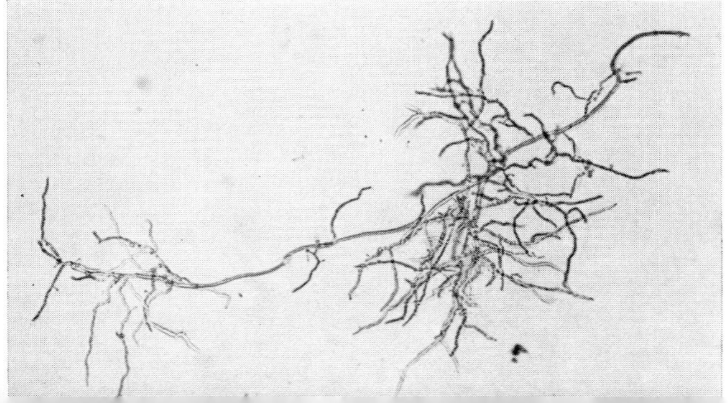

Plate 10. Part of a thin soil section showing quartz particles with a coating of humus. (× 125)

Plate 11. Fungal hypha extracted from soil and placed on a nutrient medium. The branching hyphae have grown out from the central straighter one after a period of incubation. (× 125)

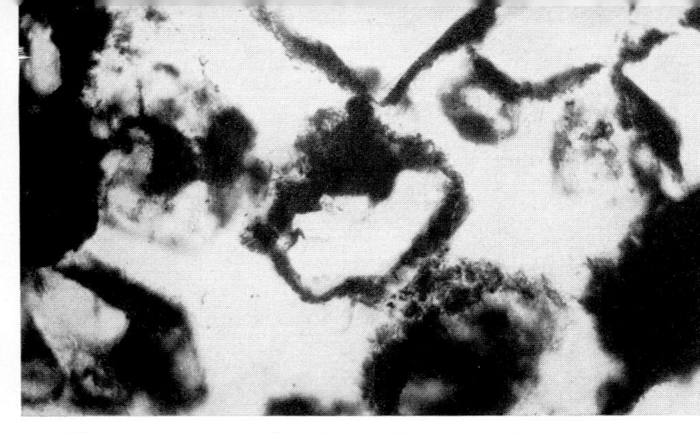

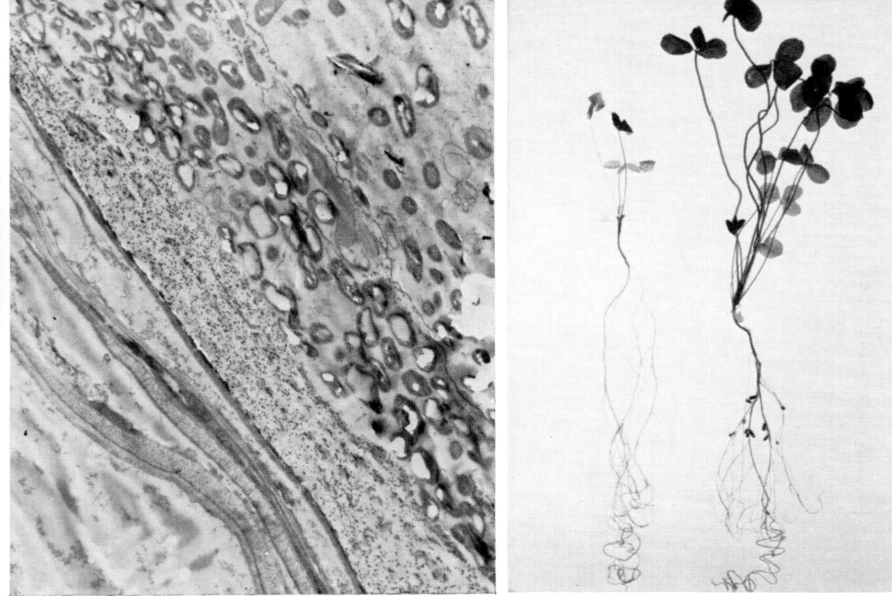

Plate 12. Electron microscope photograph of section of legume root inoculated with nodule bacteria. To the lower left are root cells. Immediately outside these is a granular layer of unknown nature, and outside this are oval cells of *Rhizobium*. (× 6,000) (By permission of Dr P. J. Dart, 1964, *Arkiv. für Mikrobiologie*, 47, 344–78.)

Plate 13. Red clover seedling inoculated with effective Rhizobium Strain (right). Note root nodules and vigorous growth. The seedling on the left is not nodulated. (By permission of Dr P. S. Nutmen, 1949, *Heredity*, 3, 263–91.)

3.2.2 *Dissection and sieving*

Careful dissection of clods removed from the soil mass with the minimum of disturbance and thorough examination with a hand lens or low-power microscope, make it possible to remove many soil animals for further study, and also to observe some microorganisms, such as fungi and even actinomycetes. The vegetative parts of micro-fungi, individual hyphae, and sometimes hyphae aggregated into strands (rhizomorphs) or compact, resting bodies (sclerotia) can nearly always be found. Fungal fruiting bodies are less often seen, but may occur in small pockets of air, particularly on decomposing plant remains.

Larger animals can be collected from the litter layers of woods and forests by using a modified sweep net. The net is made of stout cotton fabric and has a wire sieve of about $\frac{1}{3}$-in. mesh inserted two-thirds of the way down. The hole at the bottom of the net is threaded with a string so that a 3 in. by 1 in. specimen tube can be tied in it. A few handfuls of leaf litter are put in the net and shaken about so that the animals fall through the sieve and are collected in the specimen tube. In this way specimens of the litter fauna can be collected quickly.

Another way is to shake the litter in a small garden sieve and let the animals fall through on to a sheet of paper. Many will remain motionless for a few seconds but when they move they can be seen easily and picked up with forceps or a moistened camel-hair brush.

The disadvantages of these methods are that soil structure is unavoidably destroyed, soil animals are disturbed by sudden exposure to light, etc., and move from their natural micro-environments, so that the exact relations of the organisms found to soil structure is lost. Also, the smaller organisms cannot be seen as observations must be made with reflected light, and high-powered microscopes cannot therefore easily be used.

3.2.3 *Soil sectioning*

The development of methods for preparing thin sections of soil provides a means of studying soil organisms *in situ* in their true relationship to soil micro-structure. The sample to be sectioned is removed to the laboratory with the least possible disturbance, rapidly frozen and then dried from the frozen state under vacuum. When the soil is quite dry, unpolymerized synthetic resin with added catalyst is slowly introduced under reduced pressure, until the soil is completely covered. The pressure is then allowed to return slowly to atmospheric pressure, by which time the soil is thoroughly impregnated with resin. When the resin-impregnated block is quite hard, slices a few millimetres thick are made with a rock-cutting saw, then ground down to the required thickness with an abrasive and the faces of the resulting section polished. Part of a typical section is shown in Plate 10. Fungi and small soil animals can be clearly seen in such sections, which can be examined under quite high magnifications. If the soil sample is stained before impregnating it with resin, bacteria can also be seen.

3.2.4 *Agar film method for examining and counting soil microbes*

Much can be learnt about the distribution of soil organisms and their relationship to soil micro-structure from the above methods, but these do not give accurate information about the numbers of organisms in soil. Bacteria and fungi have most often been counted indirectly by culturing methods that will be described later. Culturing methods have their uses, but as we shall see, also have rather serious disadvantages. If we are mainly concerned with making as complete a count as possible of all the bacteria in a sample of soil, or if we wish to calculate the quantity of fungal hyphae, this can be done very satisfactorily using the agar film method developed by P. C. T. Jones and J. E. Mollison (Fig. 3–1). The soil sample is first passed through a sieve with 2-mm holes, and 2 g then put in a small beaker or crucible; 5 ml of distilled water are added and the soil ground thoroughly with a glass rod to break up soil crumbs. The soil suspension formed is poured off into a 100-ml conical flask, the remaining sediment rewashed in another 5 ml of water and this suspension is also poured into the flask. This process is repeated if necessary until only the heavier sand fraction is

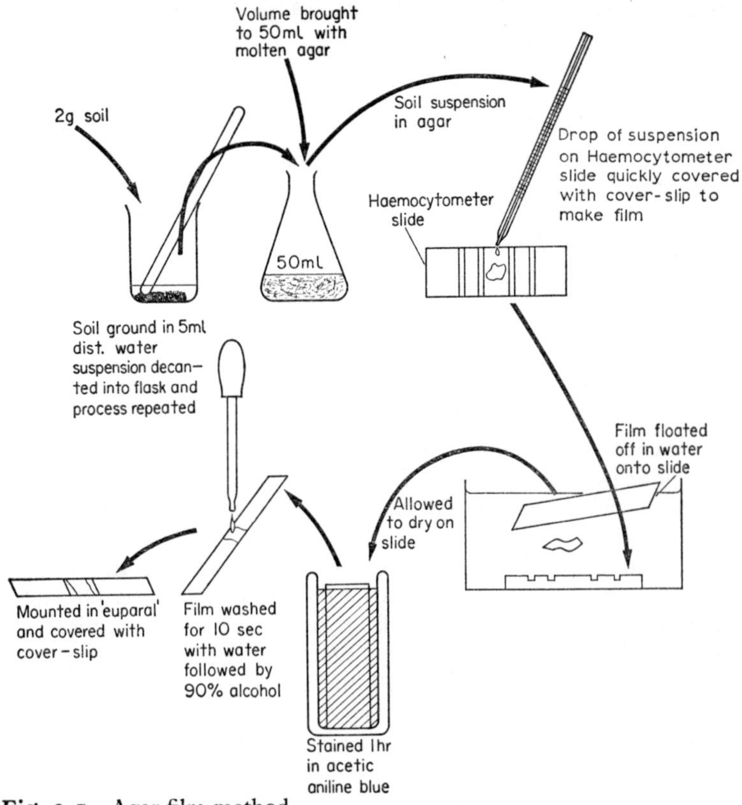

Fig. 3–1 Agar film method.

left behind. The soil suspension in the flask is then made up to the 50-ml level (already marked) with molten, 1·5% distilled water agar. The agar should have been filtered when hot through a Whatman No. 1 filter paper, and kept hot in a waterbath until used. The suspension of soil in agar is shaken vigorously, and allowed to settle for five seconds. A sample is then removed from just below the surface with a small pipette, and one or two drops placed on a clean and grease-free haemocytometer slide 0·1 mm deep. The agar is immediately covered with a cover-glass, which is gently pressed into contact with the slide to produce a film of suspension 0·1 mm thick. The slide is immersed in cold, distilled water, and in a few seconds, when the agar has set, the cover-glass is carefully slid off without damaging the agar film that remains on the slide. Surplus agar is removed with a sharp scalpel from the moat of the haemocytometer slide, and the agar film very gently floated off, first into the water and then on to an ordinary clean microscope slide. If it is difficult to get the film off the haemocytometer slide, it is probably because this was not completely clean to begin with. The film is allowed to dry slowly on the slide to which it adheres, and is then stained by immersion for 1 hour in the following solution: phenol (5·0% aqueous) 15 ml; aniline blue W.S. (1·0% aqueous) 1 ml; glacial acetic acid 4 ml, filtered about 1 hour after preparation. The films are quickly washed, first in distilled water and then in 95% ethanol, and finally mounted under a cover-slip in 'Euparal' (Raymond A. Lamb). Under a low-power × 10 objective fungal hyphae and sometimes spores can be seen in these preparations, stained blue if they were alive when the soil was sampled. Bacteria can be seen with a dry high-power objective, but are easier to distinguish with an oil-immersion lens. They usually occur either singly or in small clusters, but sometimes in quite large micro-colonies embedded in unstained mucus. If the area of the microscope field is calculated (using a micrometer slide) then the volume of original soil suspension viewed in each microscope field can be determined, and so the equivalent weight of soil. In this way, by counting all the bacteria seen in a sample of fields, the number of bacteria in a gram of the original soil can be calculated. The amount of fungal mycelium can be assessed by measuring the length or counting pieces of hyphae in a given number of fields under a lower power. It is not certain that all bacteria and fungal hyphae seen in agar film preparations were alive when the sample was taken, but bacterial and fungal cells usually lose the ability to take up stain soon after they die. Even if agar films are not used for making quantitative measurements, they provide a good means of seeing the appearance of bacteria and fungi as they have been growing in soil.

3.2.5 *Colonization of substrates introduced into soil*

The methods described make it possible to see organisms either actually in soil or in preparations made from soil, but they do not tell us very much about the organisms' nutrition. One way of finding out what types of organisms attack a particular material in soil is by burying small pieces of the material in question in soil and removing them after different lengths of time to see which organisms are present. Cellulose is especially valuable for this kind of study as it is probably the most abundant carbon compound naturally entering soil, and so the succession of organisms that attack and decompose it in different kinds of soil is of great interest. It also has the advantage that it can be easily obtained as a transparent film used for wrapping (cellophane). This is a pure form of cellulose, but it is sometimes coated with a lacquer that must first be removed in boiling water before the cellophane can be used for this purpose. It is convenient to sterilize small squares of cellophane by boiling them in water, and then to allow them to adhere to cover-glasses, which they do quite readily. The cover-glasses make it easier to handle and find the pieces of cellophane, particularly when these are partly decomposed. Cover-glasses with attached cellophane are buried in soil in a container such as a tumbler or beaker, and removed for examination after different periods of time. Organisms growing on the cellophane can be seen more easily when stained with picronigrosin in lactophenol (phenol 10 g; lactic acid 10 g; glycerol 20 g; 2·0% water-soluble nigrosin in saturated picric acid solution 10 g). Other materials than cellulose that can be used in similar experiments are chitin (present in arthropods and fungal cell walls) and keratin (present in skin, hair, feathers, etc.). If opaque substances, such as small pieces of sterilized straw, are used, direct examination after removing from soil may give little information about organisms present, and it becomes necessary to allow organisms that have colonized the material to grow out into a culture medium.

3·3 Indirect methods for microorganisms

The difficulty of observing organisms colonizing opaque substrates indicates why it may be necessary to use cultural or indirect methods to study such soil organisms as bacteria, fungi, protozoa and algae. Cultural methods are not difficult to use, and by means of them we can make counts of soil organisms. The difficulty arises when we try to interpret our results. The reasons for this will be discussed after describing the methods used.

3.3.1 *Aerobic bacteria and actinomycetes*

To count bacteria by the dilution plate method (Fig. 3–2), a convenient quantity of soil, perhaps about 10 g, is weighed and suspended in 100 ml of sterile water by shaking vigorously for a few minutes. This suspension will contain too many bacteria for direct use. It has first to be diluted sufficiently to obtain a convenient concentration of bacteria. This is done by transferring 1 ml with a sterile pipette to 9 ml of sterile water, to give a 1 in 10 dilution, and repeating the process until the required degree of dilution is obtained.

How much the suspension must be diluted depends on the kind of soil and on the nutrient medium being used; for most soils a dilution of 10^{-5} to 10^{-8} for bacteria and actinomycetes and 10^{-3} to 10^{-6} for fungi will give convenient numbers of colonies on the plates. It is always wise when making dilution plate counts from a soil for the first time to plate out several different dilutions. A small volume, usually 1 ml, is pipetted from the appropriate diluted suspension into each of three or five empty sterile Petri dishes. Sterile molten agar medium, cooled to about 45°C, is then added to each

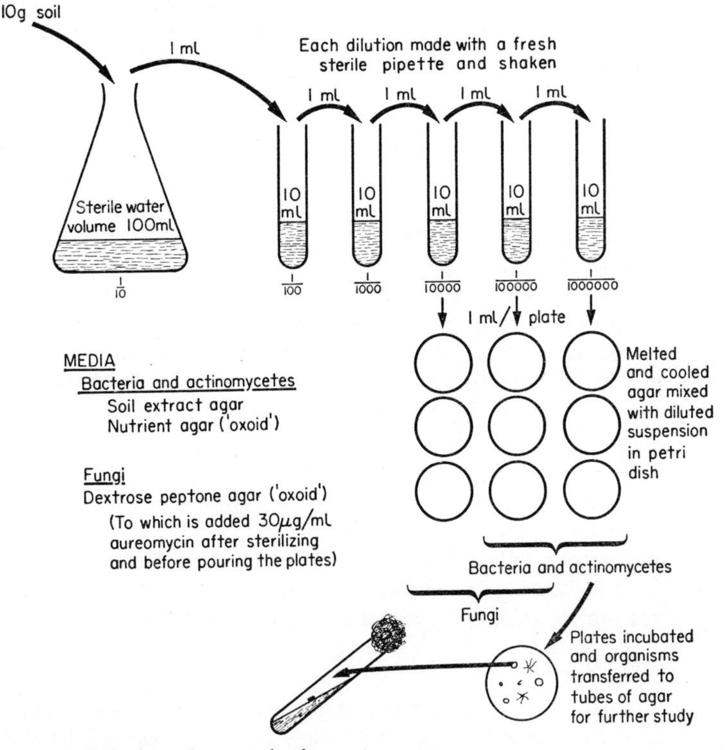

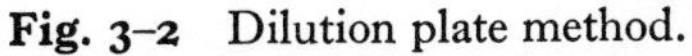

Fig. 3–2 Dilution plate method.

dish and carefully mixed with the soil suspension by rocking the dish from side to side. When the agar has solidified, the dishes are incubated, either at room temperature or in an incubator, and then examined at intervals of a few days; colonies are marked on the bottom of the dish with a grease pencil and counted as they appear. Any colonies to be studied further can be subcultured by transferring cells on a sterile wire loop to fresh sterile medium in a Petri dish or tube. Bacterial spores in soil may be counted in a similar way if the initial soil suspension is pasteurized (see 2.1.2).

The first problem when using the dilution plate method to count a large group of organisms like bacteria is to decide what medium to use.

Experience shows that largest counts of bacteria or fungi can be obtained on nutritionally weak media, for example, one containing only aqueous soil extract and a small quantity of phosphate. Some organisms grow rather slowly on a medium like this, so it is necessary to incubate the plates for quite long periods, up to about five weeks, to obtain the maximum count. Ordinary room temperatures are suitable for incubation. Although many kinds of organisms will grow on a weak medium like soil extract agar, there are others with specialized nutrient requirements that will not be able to develop unless these are satisfied. Also, organisms compete for nutrients and antagonize one another by the release of antibiotics into the medium; this prevents those organisms that are poor competitors from developing. These effects will be less, the less crowded the colonies are on the dilution plates. Soil bacteria present in very small numbers will almost inevitably be overgrown by more abundant species unless the medium specially favours the less common kind.

Comparison between counts of bacteria made directly by the agar film method and indirectly from dilution plates shows that only a small part of the total bacterial population is counted by dilution plates. For example, counts of bacteria made on dilution plates in two kinds of soil at Rothamsted were approximately $\frac{1}{10}$ to $\frac{1}{20}$ of those made from the same sample using the agar film method. This is because only a small proportion of soil bacteria can grow even on a non-selective medium like soil extract agar, and also because bacteria in micro-colonies that are easily counted on agar film preparations may not be separated into individual cells when soil suspensions are prepared. The result is that one colony on a dilution plate may in fact arise from a group of cells and so give too small a count.

Actinomycetes can often be seen and counted on the same medium that is used to count other bacteria, but media more favourable to these organisms can be used, for example one containing chitin.

3.3.2 *Anaerobic bacteria*

Cultures that are incubated in the atmosphere, without taking any special precautions to exclude oxygen, will allow only aerobes and facultative anaerobes to develop. Important kinds of soil bacteria are, however, obligately anaerobic and can grow only in the absence of oxygen. To isolate these bacteria from soil, it is necessary to incubate the isolation plates in an oxygen-free atmosphere. There are various ways of doing this, such as using gas-tight jars from which oxygen is removed by flushing out with another gas or using chemicals that absorb oxygen. A simple alternative is to make use of the respiration of other aerobic organisms or tissue to remove oxygen from a closed system, so producing anaerobic conditions. Pure cultures of aerobic organisms can be used for this purpose but are not necessary. Good results can be obtained in the following way (Fig. 3–3). Use either a glass cylinder about 6 in. in diameter and 12 in. to 18 in. high that can be sealed with plasticine and a glass plate, or a small desiccator as a gas-tight container. If a cylinder is used, a tumbler is inverted in the bottom to act as a stand for the isolation plates. Oats or other grain are put in the bottom of the cylinder or desiccator, to occupy at least a tenth of the total volume, and thoroughly moistened with tap water. Plates to be incubated anaerobically are then placed in the container, which is carefully sealed to make it gas-tight. The whole container is then incubated and, as the grain starts

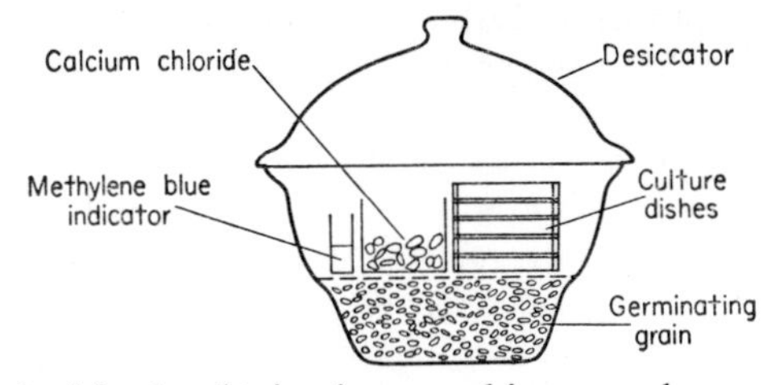

Fig. 3–3 Method for incubating in anaerobic atmosphere.

to germinate, it respires actively, using up the available oxygen and so creating anaerobic conditions. The carbon dioxide produced in respiration stimulates growth of some anaerobic bacteria. Condensation within the culture plates may be troublesome and this may be diminished by placing a small open dish of calcium chloride at the top of the pile of plates. It is important to have some means of checking that anaerobic conditions have definitely been established and that there is no leak in the system. This can be done with a chemical indicator, such as methylene blue, which is coloured in the oxidized state but not when reduced. Methylene blue indicator for this purpose can be prepared as follows: make up three stock solutions: (1) 6 ml 10N NaOH diluted to 100 ml with distilled water; (2) 3 ml of 0·5% aqueous methylene blue diluted to 100 ml with distilled water; (3) 6 g glucose in 100 ml of distilled water, to which has been added a crystal of thymol. Each time the indicator is used, equal parts of the three solutions are mixed together in a test tube and boiled until the methylene blue is decolourized. If the decolourized indicator is now placed immediately in the container the blue colour will return, but should disappear within a day or two, as anaerobic conditions are established. Opening the container to examine the cultures admits oxygen and some time will be needed to re-establish anaerobic conditions.

3·3·3 *Selective media*

Incubating anaerobically is one way of making the isolation process selective for particular organisms. Another is to use a medium that favours only certain organisms. A good example of such a medium is one that contains no compounds of nitrogen. Most organisms require such compounds and will not therefore be able to grow on the medium, but those that can fix atmospheric nitrogen will have an obvious advantage, and provided all other necessary nutrients are present will be able to grow well. It is difficult to make a solid medium entirely free from nitrogenous compounds, but organisms that cannot fix nitrogen will grow only very feebly, whereas active nitrogen-fixing organisms like *Azotobacter* will grow much more vigorously and quickly.

3.3.4 *Selective temperatures*

Conditions of culture can also be made selective by changing the temperature of incubation. Most soil organisms grow well between 20°C and 30°C and these are described as mesophilic, but some bacteria and actinomycetes, and a few fungi, grow well at temperatures up to 65°C. These are thermophilic organisms; some are facultative thermophils and can grow at lower temperatures, but others are obligate thermophils and cannot grow at normal temperatures. Thermophilic organisms are common in some soils, but are especially abundant in compost and silage heaps, which ferment and become hot. *Thermopolyspora polyspora*, a thermophilic actinomycete, is common on hay that has heated, and can cause the disease known as 'farmer's lung', produced in some people who inhale its spores in the dust of mouldy hay. It grows best at 50°C, and does not grow either below 35°C, or above 65°C. Microorganisms that grow best in the cold, psychrophils, are rarer than thermophils. An example is a yeast, *Candida scottii*, isolated from antarctic soil. Strains of this yeast have been found that grow as well at 0°C as at 10°C, but not at all at 15°C or higher.

3.3.5 *Fungi*

Similar methods to those used to count and isolate soil bacteria and actinomycetes can be used for fungi. Bacteria growing in a medium can interfere with the growth of fungi, so it is usual to incorporate in the medium an antibiotic, such as aureomycin, which is active against bacteria but not fungi. Dilution plate counts can tell something about soil fungi, but can also be very misleading. This is because nearly all the colonies that develop on a soil dilution plate come from fungal spores and not hyphae, so counting colonies tells us only how many spores are in the soil and gives no indication of which fungi are active. When the spores of a particular fungus are abundant in the soil it is fairly certain that that fungus was growing actively at some previous time, but we can easily get an unbalanced picture, even about earlier activity, because some fungi produce very many spores, while others which may have just as much mycelium, produce few. The value of dilution plate counts for estimating fungal activity in soil is still further limited by the fact that the endophytic mycorrhizal fungi will not usually grow in culture.

A method for determining which types of fungi are active is to remove hyphae from soil and put them on to a nutrient medium where they can develop and give rise to colonies that can be identified (Plate 11). This can be done by picking hyphae individually with fine forceps or a needle from a soil suspension under a low-power microscope, or by separating hyphae from soil by a process of wet sieving and sedimenting, and then plating out the resulting suspension that consists largely of hyphae. With either method it is necessary to examine the hyphae periodically after plating out to see which have started to grow; these can then be removed to a tube of sterile agar medium. This method has isolated from soil many fungi, particularly Basidiomycetes, that never appear on dilution plates, probably because they are always overgrown by vigorous, sporing species. Many of the hyphae extracted from soil may not grow, either because they are dead, or because they are of species that cannot be cultured.

3·3·6 *Washing methods*

The methods described so far do not tell us which microorganisms are growing on which type of material in soil. It is possible to find out something about this by removing pieces of identifiable material, perhaps root or dead leaf, washing very thoroughly in several changes of sterile water

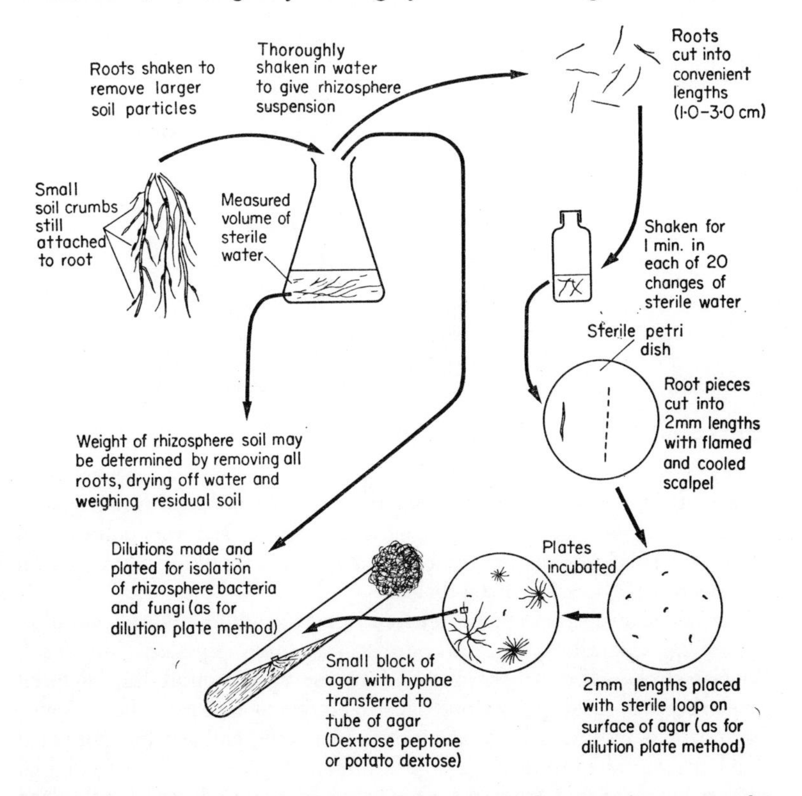

Fig. 3–4 Isolation of organisms from rhizosphere and root surface (rhizoplane).

to remove any spores on the surface of the material, and then plating out on a suitable selective or non-selective agar medium. This method is much used for studying rhizosphere and root-surface organisms (Fig. 3–4).

3·3·7 *Protozoa*

So far we have discussed methods for counting and isolating organisms that produce easily visible colonies on nutrient media. Protozoa do not do this; although they often occur on bacterial isolation plates, they rarely develop into obvious colonies and cannot be counted reliably on the plates, so special methods are needed for their study. Soil protozoa feed on bacteria, but only some species are edible; others are inedible and some are actually toxic to protozoa. On bacterial isolation plates it is a matter of chance whether a protozoan finds itself amongst edible bacteria where it can develop, or inedible bacteria where it cannot. B. N. Singh developed a method (Fig. 3–5) that overcomes these difficulties; instead of leaving the source of food to chance he provided known edible bacteria.

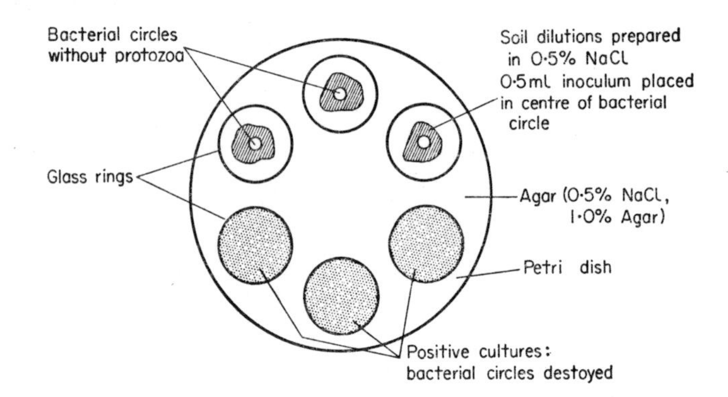

Fig. 3–5 Method for counting protozoa.

In his method a culture of edible bacteria, such as *Aerobacter* is first grown on a medium giving good growth. Cells are removed from this culture and spread in small circles on the surface of water agar (containing 0·5% sodium chloride and 1% agar) in Petri dishes.

A soil dilution series is then prepared in 0·5% sodium chloride solution and 0·5 ml quantities of each dilution of these pipetted into each bacterial circle. Glass rings pressed into the agar around the bacterial circle prevent spreading from one bacterial circle to another. The presence of protozoa is indicated by clearing of the bacteria, and can be confirmed by microscopical examination. The number of protozoa in the original soil can be calculated from the proportion of bacterial circles colonized by protozoa at the different dilutions.

3.4 Indirect methods for soil animals

Because of the enormous range in size of soil animals, from larval mites no bigger than some protozoa to earthworms several inches long, and because of differences in their biology, structure and behaviour, no one method can be used successfully to collect or sample all kinds of animals. Special methods must be applied for specific groups and because each method has its own advantages and limitations care is needed when comparing results from different methods.

Some methods, originally used simply to collect soil animals, have been suitably refined to give quantitative estimates. Similarly, quantitative methods can sometimes be simplified for qualitative studies. But it is always extremely difficult to test the efficiency of any particular method for collecting or extracting animals from soil samples. Often all that can be done is to compare results got by different methods and select the best. Much of soil zoology has been concerned with improving methods for estimating the abundance of soil animals, and the better methods have revealed unexpected organisms in unsuspected numbers. For example, the Protura, a primitive order of insects, were first described by an Italian entomologist, Sylvestri, in 1907, and were thought to be rather uncommon (many entomologists have never seen one); yet, when an accurate flotation method was used, more than 250 specimens were recovered from a single soil sample 2-in. diameter and 9-in. deep taken from a grass plot at Rothamsted.

3.4.1 *Berlese and Tullgren Funnels*

The apparatus that Berlese described was intended simply as a means of collecting specimens from soil, leaf litter, vegetable refuse, etc., for systematic work. The material was held in a wire tray in the top of the funnel-shaped water jacket (Fig. 3–6a) and specimens fell through the wire tray into the funnel and then into the collecting jar when the water was heated and dried out the soil or litter. Tullgren dispensed with the water jacket and dried the soil or litter with heat from an electric light bulb suspended in a metal cylinder above the funnel and his 'Tullgren Funnel' has been much modified since. All are based on the principle that the animals move downwards to the bottom of the soil sample as it dries out and, finally, fall into the funnel and collecting vessel. Heat helps to drive the animals downwards and in most Tullgren Funnels the soil surface nearest the heat source is about 30°C. Light from the bulb is thought to have little effect. Understandably, the performance of Tullgren Funnels is affected by many factors such as the form and moisture content of the sample, the rate at which the sample is heated and dried, and the reactions, particularly to temperature and moisture, of the many different species in the sample. No specific design of Tullgren Funnel can be expected to extract all kinds of organisms with equal efficiency. Almost every

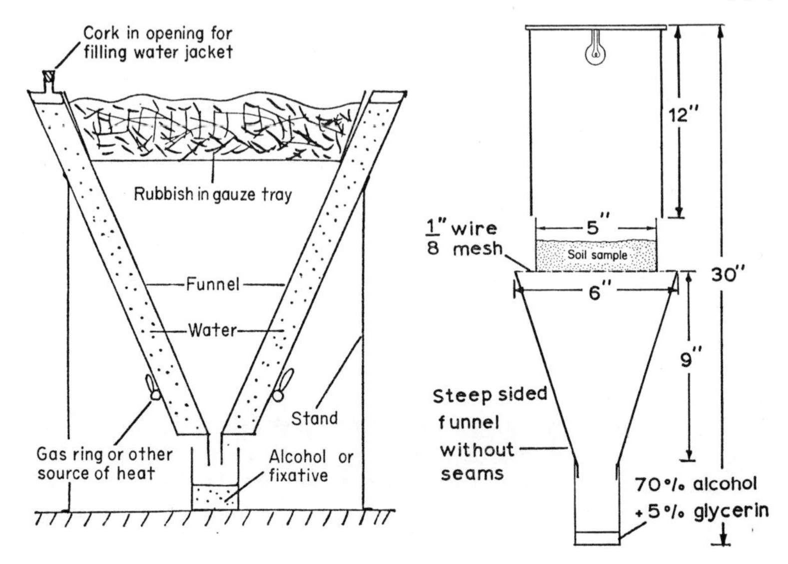

Fig. 3–6a & b Diagrammatic sections of (a) Berlese Funnel and (b) Tullgren Funnel. (**3–6a** from Brit. Mus. (Nat. Hist.) Instructions for Collectors, No. 4a.)

worker has produced a new design or modification and the best that can be done here is to illustrate an apparatus that is simple to construct and has been found to work well for different soils and different organisms.

The apparatus (Fig. 3–6b) was designed to take a soil sample 2-in. diameter and 6-in. deep (i.e. about 20 cu. in. of soil) that was crumbled before being put on the sieve. The 25-watt bulb is placed so that the temperature at the soil surface is approximately 30°C. Extraction is complete in 3–4 days but may take longer with wet samples.

3·4·2 *Flotation methods*

Berlese and Tullgren Funnels depend on the animals being stimulated to leave the sample and their performance and efficiency differs greatly for different animal groups and species. Moreover, they cannot extract resting stages such as eggs and pupae. These disadvantages do not apply to flotation methods that depend only on physical principles for separating the animals from the soil and litter. One such method was developed during the Second World War to meet the need for estimating wireworm populations accurately. It has become a standard method for extracting arthropods from soil samples and can be modified for special requirements.

The apparatus (Fig. 3–7) was designed to take a soil sample 4-in. diameter and 6-in. deep (i.e. about 80 cu. in. of soil). The mesh size of

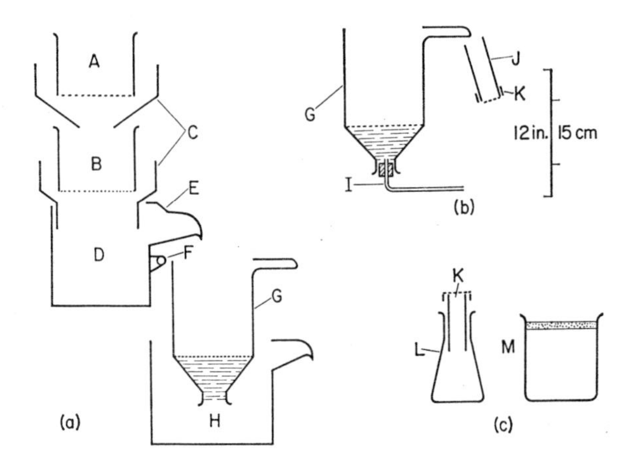

Fig. 3–7 Flotation method for soil arthropods. (A) 5 mm mesh sieve, (B) 2 mm mesh sieve, (C) splash guards, (D) reservoir with deep lip (E) and pivot (F), (G) flotation vessel with fine sieve (70–100 mesh), (H) tank receiving effluent from G, (I) connection to compressed air supply (J) glass cylinder, (K) bolting silk sieve (70–100 mesh), (L) 750 ml wide-necked flask, (M) 1000 ml beaker for xylene/water separation stage.

the sieve in vessel *G* determines the size of the organisms extracted. After pretreating the sample the extraction process is done in three stages.

PRE-TREATMENT. The sample is crumbled into a suitable container, covered with water, frozen for a few days and then thawed. The freezing and thawing breaks down many of the soil crumbs and makes it easier to extract the arthropods. Another advantage is that samples can be stored frozen for several months, without deteriorating, if they cannot be extracted straight away. When the soil cannot be frozen the soil crumbs can be dispersed to some extent by soaking the sample in a solution of 50 g sodium hexametaphosphate and 20 g sodium carbonate per litre. Chemical dispersion can be helped by withdrawing the air from the sample in a vacuum desiccator after the sample is immersed in the dispersion solution.

FIRST STAGE. The object of the first stage is to wash the arthropods free from vegetation, such as grass roots, and to break up the soil crumbs so that animals are not trapped in them. Soil crumbs that pass a 2 mm sieve rarely contain animals over 1 mm long and even for much smaller animals little is gained by breaking the soil into smaller crumbs. As the sample is washed through sieve *A*, stones are removed and vegetation, such as grass roots, is discarded after being carefully teased apart under the water jet to wash out any animals. Sieve *A* and its surrounding splash guards are then removed and the process repeated for sieve *B*. Sieve *B* is then inverted and its remaining contents washed into the reservoir *D*, which has held back the bulk of the sample during the washing process so that the fine sieve *G* is not blocked. The contents of the reservoir *D* are then washed into the vessel *G*. During washing, much of the silt and clay in the sample passes through the sieve in *G* and so decreases the bulk of the sample.

SECOND STAGE. The object of the second stage is to separate the organic material, including the soil animals, from the mineral fraction of the sample. Vessel *G* is placed in a stand and the opening at *X* is connected to a compressed air supply—which can be improvised by using a motor-car tube as a compressed air supply. *G* is two-thirds filled with magnesium sulphate solution of specific gravity 1·2 and the mixture thoroughly agitated by stirring and by blowing a stream of compressed air through the sieve from below. When the mixture is allowed to settle, the buoyant material, including the soil animals, floats and is collected by adding more magnesium sulphate solution and tilting the vessel so that the floating material passes over into the sieve *J*, which is a glass cylinder closed with bolting silk of the same mesh as the sieve in *G*. The magnesium sulphate solution can be recovered and used repeatedly, provided the specific gravity is maintained.

THIRD STAGE. At this stage the arthropods are separated from the organic material. This depends on the fact that the arthropod cuticle is wetted by some compounds, such as paraffin oil, benzene and xylene, but not by water. The material extracted by flotation is washed thoroughly in sieve *J* and then washed into a wide-necked flask. About 25 ml of xylene are added, the mixture shaken well and then washed into a beaker, which is then filled up with water. The arthropods, wetted by xylene, accumulate in the xylene layer, which settles out on the water surface. The other organic material remains in the water below the xylene–water interface. The xylene–water interface is scanned under a low-power binocular mounted on a long arm stand and the arthropods picked out. A tiny wire loop mounted in a wooden handle is useful for this purpose.

The flotation method has its limitations. For example, the cuticle of some arthropods, such as cecidomyid larvae and some insect eggs, is wetted by water, so they cannot be separated by xylene from the other organic material; they must be picked out at the flotation stage, which can be repeated for this purpose in a beaker of magnesium sulphate solution. When the soil sample contains much organic matter, as with peat soils or forest litter, the flotation stage may serve little purpose and the xylene separation stage may be hampered by organic debris holding some of the arthropods below the xylene–water interface.

The apparatus described was designed to extract arthropods about 2 mm long and 0·2 mm broad. Smaller organisms, such as the immature stages of mites and collembola, will be lost through the meshes of the sieve in *G*. For such micro-arthropods a small-scale apparatus should be used. The technique is essentially the same but the samples are much smaller, usually about 50–100 cu. cm and much finer sieves are used in *G* and *J* to retain the micro-arthropods.

3·4·3 *Earthworms*

Estimating earthworm populations presents special problems because some species such as *Lumbricus terrestris* can burrow several feet deep and it is impractical to take soil samples to that depth. Moreover, although earthworms are abundant in many habitats, large soil samples are needed to ensure a reasonable number of earthworms per sample. For the smaller species, such as *Eisenia rosea*, *Lumbricus castaneus*, and *Allolobophora caliginosa*, which live in the top-soil, a good estimate of the population can be got by hand sorting samples about 1 ft square and 8–12 in. deep. However, this method is laborious and for qualitative work, or for quantitative estimates of species that burrow deeply, other methods are preferable. One way is to apply a liquid that penetrates the earthworm burrows and stimulates the worms to come to the soil surface. Dilute formalin is the most effective and is used as follows: 25 ml of 40% formalin is added to 1 gal of water and the solution applied to an area 2 ft square. Worms begin to come to the surface after a few minutes but those in deep burrows take longer. A second application of formalin solution is given when the first no longer brings worms to the surface.

The technique works best when the soil has not been recently disturbed and its temperature is about 10°C. Then it gives an accurate estimate of species like *L. terrestris*, which have a well-defined burrow system. Only a proportion of the other species are recovered and the best estimate of the total earthworm population is got by using hand sorting for the species that live in the top-soil and the formalin method for those that live deeper.

Worms expelled by formalin survive well if they are immediately rinsed in water and the method is excellent for collecting specimens for laboratory work.

Although earthworm cocoons can be found by hand sorting, it is quicker and more accurate to use a flotation method. A simple method is to gently wash and shake a soil sample in an 8 mm mesh sieve suspended inside a 1 mm mesh sieve standing in a bowl of water (the largest cocoons will pass an 8 mm sieve and the smallest will be retained by a 1 mm sieve). After the soil is washed through the 8 mm sieve, the 1 mm sieve is shaken in the water for a time to let silt pass through. Then it is immersed in a bowl of magnesium sulphate solution of specific gravity 1·2. The cocoons rise to the surface and can be picked out.

3·4·4 *Nematodes and enchytraeids*

In Berlese and Tullgren Funnels the soil or litter is dried and the animals leave the sample to escape desiccation. This technique is suitable for animals that live in the air spaces in the soil and those that can tolerate lack of water or some fall in humidity. It cannot be used successfully for the animals that live in the water films surrounding the soil particles or those that are particularly susceptible to desiccation. For them, an extraction method in water is needed. Some organisms, such as the *Enchytraeidae*,

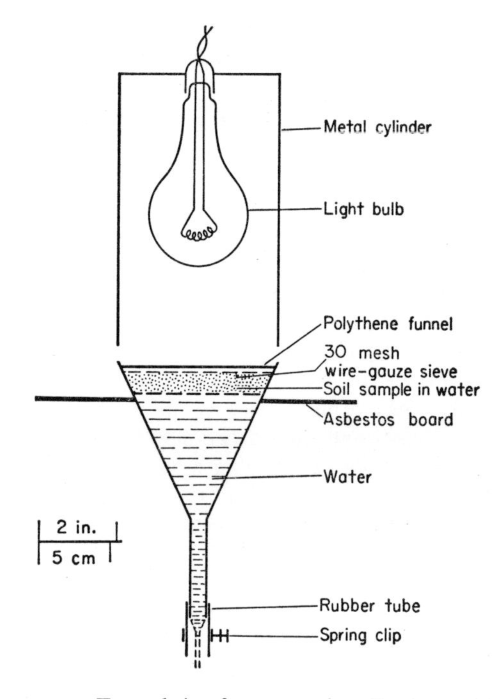

Fig. 3–8 Baermann Funnel (as for extracting Enchytraeidae). (Courtesy F. B. O'Connor.)

are intermediate; specimens can be collected with a Tullgren Funnel but usually a wet extraction process is better.

One of the standard methods for collecting nematodes from soil or plant material is based on the method originally devised by Baermann for extracting hookworm larvae from soil. The apparatus (Fig. 3–8), consists of a funnel with a piece of rubber tubing attached to the stem and closed with a Hoffmann or spring clip. The funnel is held in a support and almost filled with water. The soil is crumbled on to a paper handkerchief which is then placed in a 2½–3-in. diameter sieve made by soldering 30 mesh copper gauze on to a short length of copper tubing. The sieve is placed in the funnel so that the sample is immersed in the water. Nematodes leave the soil, pass through the paper handkerchief, and, being heavier than water, sink to the bottom of the funnel stem. After some hours, or overnight, they can be withdrawn by releasing the spring clip. If they are left in the funnel too long the nematodes may die from lack of oxygen. To avoid this, the sieve containing the sample can be supported in a petri dish of water so that the sample is partially immersed and the sieve is held clear of the bottom of the dish.

Samples consisting of plant material should be cut into small pieces or macerated in a domestic mincer before being extracted. Mincing also allows nematodes to be collected from inside plant tissues. It breaks up most of the adults but not the larvae. Not all nematodes present in soil or plant samples will be extracted by the Baermann Funnel method, but much more complicated methods would be needed to do this.

By a simple modification, the Baermann Funnel apparatus can be used to extract *Enchytraeidae* from soil samples. Enchytraeids are usually much larger than nematodes so the soil sample is placed directly on to the 30 mesh sieve and immersed in the funnel. A 60-watt electric light bulb in a metal cylinder (e.g. an empty tin) is suspended above the funnel so that the temperature of the sample rises to 45°C in about three hours. The enchytraeids, which leave the sample and collect in the funnel stem, are withdrawn by releasing the spring clip.

The cysts of cyst-forming nematodes, such as *Heterodera* spp., can be recovered from soil by a method that depends on the dry cysts floating on water while the heavier soil particles sink. The soil sample is slowly air dried and crumbled and turned daily to prevent hard lumps forming. Because cysts of many *Heterodera* spp. cannot withstand complete or prolonged drying, the sample should be only partially dried when the cysts are needed for further study. It can be completely dried when only an estimate of the *number* of cysts is needed. The dried or partially dried soil is passed through a $\frac{1}{4}$-in. mesh sieve and thoroughly mixed. Thorough

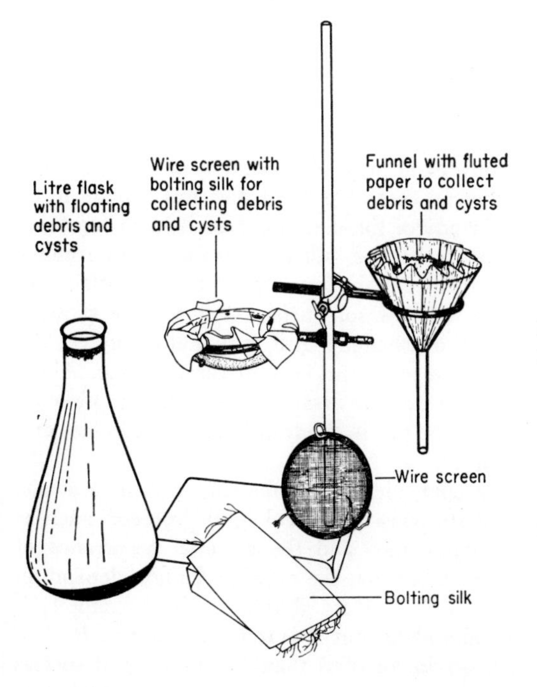

Fig. 3–9 Nematode cyst flotation method.

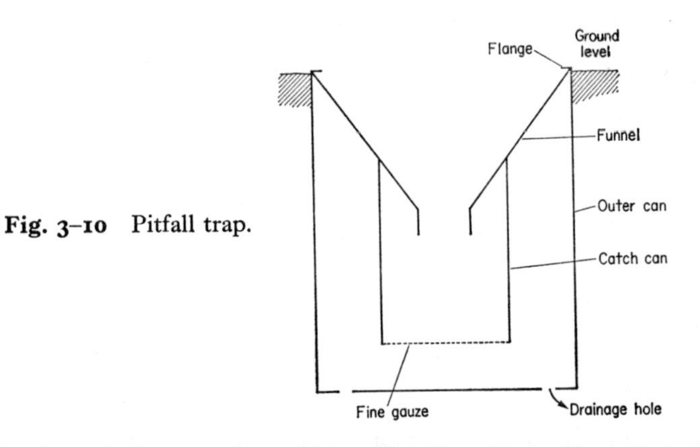

Fig. 3–10 Pitfall trap.

mixing is especially important when the original sample is got by combining many small samples. A sample of the mixed soil (up to 50 g) is weighed out into a 1-litre conical flask, which is then half filled with water and shaken vigorously. It is then filled and allowed to stand until the water in the neck of the flask is clear. The cysts and light debris that float are decanted into a pair of 20 and 60 mesh sieves nested together, and then washed in a stream of water. The coarse debris on the 20 mesh sieve is discarded and the material on the 60 mesh sieve, containing the cysts, is washed into a funnel with a fluted filter paper. When the funnel drains the cysts can be picked off the filter paper (Fig. 3–9).

3·4·5 *Pitfall traps*

Pitfall traps provide a convenient way of collecting or studying soil animals that are active on the soil surface but not plentiful enough to be collected or sampled easily by any of the methods described so far. This applies particularly to predatory beetles, such as carabids and staphylinids, or to millipedes and centipedes. In its simplest form a pitfall trap can be a tin or jar sunk flush with the soil surface. Figure 3–10 shows a more complicated trap with an overhanging rim, a funnel to protect the catch from birds, and a fine gauze at the bottom for drainage. Alternatively, rain can be excluded by a cover arranged above the trap like an umbrella and a preservative can be put in the trap to kill the captives and prevent them eating one another.

Pitfall traps are not only convenient for collecting certain animals, they provide a useful way of studying some ecological problems such as the spatial distribution of populations and the seasonal occurrence of species. It is important to remember that the numbers caught depend on the activity of the species as well as their abundance. For example, if species *A* and *B* are equally abundant, but *A* is more active than *B*, then pitfall traps will catch more specimens of *A* than *B*. Similarly, if species

C is more active in warm weather than in cold, this will be reflected in the numbers caught on warm and cold days. However, even if trap catches were not affected by activity, it is necessary to know what area the catch is drawn from, or what proportion of the total population is caught, before one can estimate the total population or the numbers per unit area.

3·4·6 *Population estimates*

An estimate of the total population can sometimes be made by the marking and recapture method. Individuals caught on a particular occasion are marked, released and allowed to redistribute themselves in the population before a second catch is made. Then, if n_1 individuals are marked and released and if the second catch, n_2, contains n_3 marked individuals, the estimated total population is given by

$$N = \frac{n_1 \times n_2}{n_3}$$

The method depends on certain assumptions that must be satisfied for the estimate to be valid. These are that the catches are random samples of the total population and that marked and unmarked individuals can be caught with equal ease. In practice it is often difficult to satisfy these assumptions, or even test that they are satisfied. For example, with pitfall traps, it is difficult to decide where to release the marked individuals and to be sure that they become redistributed in the population the traps are sampling before the second catch is made.

The method is of somewhat limited use for soil population studies, not only because it is difficult to satisfy the basic assumptions, but because many kinds of soil animals, e.g. slugs, cannot be marked satisfactorily. It is probably best used to study species of carabid and staphylinid beetles that are active, range widely, and can be marked easily, e.g. with spots of cellulose paint on the elytra.

3·5 **Culture methods for soil animals**

Most of the larger soil animals are easy to rear or maintain in laboratory cultures. Usually all that is needed is a suitable container of moist soil, enough food and, preferably, a cool place such as a cellar for storage. For earthworms and slugs, glazed earthenware pots about 8-in. diameter and 12-in. deep, firmly covered with butter muslin or perforated polythene sheeting to prevent specimens escaping but to allow ventilation, are ideal, but anything from a jam jar upwards can be used. Earthworms are best kept in friable loam; slugs in a 4 : 1 loam and peat mixture. Earthworm cocoons and slug eggs can be incubated on wet filter paper in Petri dishes. Cultures of millipedes, centipedes and woodlice can be maintained in leaf litter overlying moist soil.

For smaller animals, such as symphyla, mites and collembola, mass cultures can be maintained in screw-topped kitchen storage jars containing a moist, finely divided, soil/organic matter mixture (e.g. soil and farmyard manure or leaf mould screened to pass a 1-mm sieve) overlying a layer of plaster of Paris that is moistened periodically to maintain a saturated atmosphere. A glass cylinder embedded in a tray of plaster of Paris can be watered from outside and the culture is not disturbed.

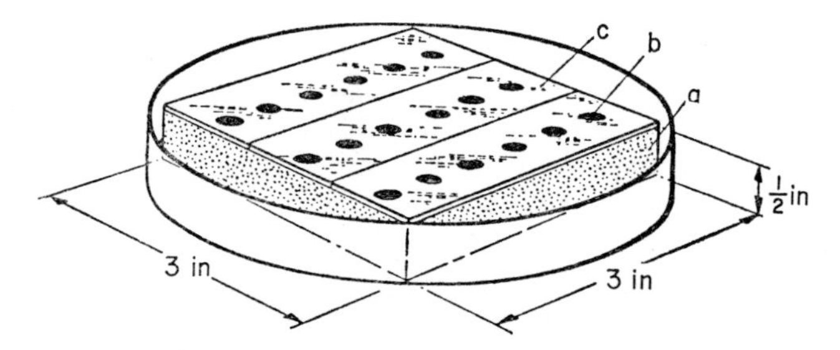

Fig. 3–11 Individual culture method for small arthropods. Petri dish with (a) plaster of Paris block, (b) $\frac{1}{4}$-in. holes, (c) covered with 3 × 1-in. microscope slides. (From *Soil Zoology*.)

Figure 3–11 shows a convenient way for rearing small arthropods individually or in small numbers. Pour some plaster of Paris into a dish and press a row of 3-in. by 1-in. microscope slides into the surface before it sets. Remove the slides and drill rows of holes about $\frac{1}{2}$-in. deep and $\frac{1}{2}$-in. diameter so that each row can be covered with a slide. The whole block can be kept uniformly moist by watering the plaster of Paris.

3.6 Soil respiration

None of the methods described so far for studying soil organisms provides a picture of the activity of all the organisms in a particular soil or horizon. One way in which this may be done is by measuring the quantity of oxygen taken up or of carbon dioxide released by a sample of soil over a period of time. The value obtained is often called the respiratory rate of the soil, but is, of course, a measure of the total respiration of all the organisms in the soil sample. Because the respiratory rate of individual organisms may vary widely with their metabolic activity, so the respiratory rate of the soil is a poor guide to the numbers of organisms contained in it, but gives a good indication of total activity. In most soils a large proportion of the total respiration is probably by microorganisms.

Soil organisms and nutrient cycles 4

4.1 Circulation of nutrients

Growth of all natural vegetation depends on the circulation of the major nutrients, which reach the soil as organic material and then undergo an often complex series of changes that restores them to forms in which plants can use them again. Man has interfered with the natural circulation of nutrients by growing and removing crops from the land and he has thus found it necessary to make regular applications of organic manures and inorganic fertilizers to maintain crop growth.

Of the elements required by plants in the largest quantities, hydrogen and oxygen are usually abundant in the environment. Shortage of carbon would soon limit plant growth if it were not naturally recirculated, and nitrogen, which also goes through a complex cycle, soon becomes deficient in soil that is regularly cropped unless replaced by nitrogenous fertilizers. Phosphorus and sulphur, although needed by plants in smaller quantities than nitrogen, are just as essential and also pass through complex cycles.

4.2 The carbon cycle

It is convenient to start discussing the carbon cycle (Fig. 4–1) with the green plant. This obtains its carbon from atmospheric carbon dioxide, which it fixes and converts into sugar by the process of photosynthesis. Carbon dioxide is usually present in the atmosphere at a concentration of about 300 p.p.m., but near growing plants it may be less than 300 p.p.m., and this will slow down the rate of photosynthesis. For this reason, the carbon dioxide concentration in greenhouses is now sometimes artificially increased to obtain bigger yields of crops like lettuces and tomatoes.

Some carbon is returned directly to the atmosphere by the respiration of plants; some is excreted by the root into the soil in the form of sugars, amino acids and other organic compounds, but most is incorporated in the plant body, either as structural material like cellulose and lignin, or in the many compounds that constitute protoplasm and storage material. All parts of plants eventually die and decompose, but the way in which this happens depends on the type of plant. Annuals die completely within a year, and parts of perennial plants are constantly dying and being replaced. Small roots have a limited life and leaves and often branches are regularly shed. The quantities of plant tissue returned to the soil by perennial plants can be judged from calculations made for a tropical forest; that in every acre about 5,000 lb of carbon are deposited as leaf and branch litter every year, and about 3,000 lb of carbon are added to the soil as dead roots.

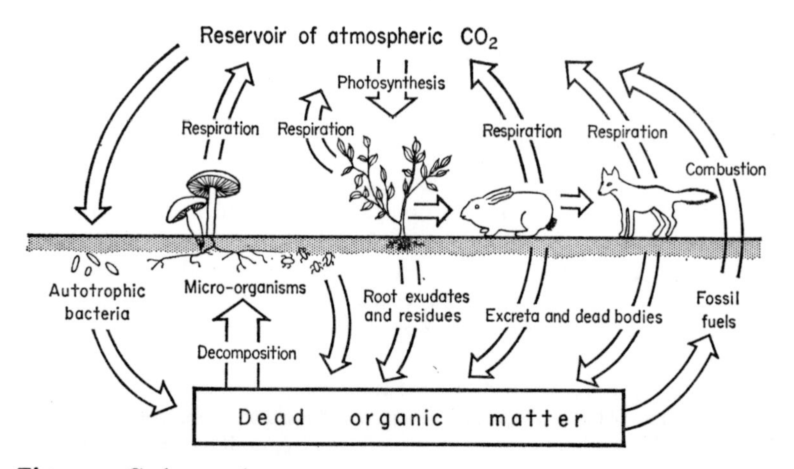

Fig. 4–1 Carbon cycle.

4.2.1 *Decomposition of carbon compounds*

In a situation in equilibrium, when organic matter is not accumulating, decomposition must be fast enough to remove the equivalent of all the organic matter being added to the soil, and to replenish the supply of atmospheric carbon dioxide. The stages through which the carbon passes before being finally converted into carbon dioxide may be many and complex. As carbon passes from one organism to another, part of the total is released as carbon dioxide, and the remainder incorporated into new microbial tissue, or is converted into an organic residue highly resistant to further decomposition. The plant constituents most rapidly decomposed in soil are carbohydrates, lipids and proteins, all of which provide substances readily used by many kinds of organisms. When substances of this type are present in soil, other nutrients are not usually lacking. The constituents of the plant framework, cellulose, lignin, suberin, etc., decompose more slowly. They may be attacked first by fungi or bacteria, but are often eaten by soil animals. These digest part of the plant material and excrete a residue that forms food for other organisms. This preliminary action by soil animals often increases the rate at which subsequent microbial decomposition occurs. This may be partly because the material is chewed into small fragments, increasing greatly the surface area available for attack by microorganisms, but also because it distributes the material more widely in the soil. Earthworms do this most effectively by dragging dead leaves down into their burrows where the environment is much more favourable for decomposition than at the soil surface (2.2.2).

Cellulose is usually decomposed faster than lignin, but when lignin-decomposing Basidiomycetes are very active, as they may be for example in beech-leaf litter, lignin may be decomposed equally rapidly. The fungi that grow on cellulose and lignin may themselves be eaten by mycophagous animals such as mites, springtails and nematodes, which when they die, provide nutrients for further bacteria or fungi. When bacteria are the primary colonizers of cellulose, they may be eaten by protozoa, which in turn are either eaten by their predators or die and their remains consumed by bacteria and fungi. All these organisms respire, and so constantly return some carbon to the atmosphere. Some of the complex organic compounds synthesized by microorganisms are highly resistant to digestion by other organisms and accumulate to form the humus fraction of the soil. Part of this fraction is always very slowly decomposing.

4.2.2 *Herbivorous animals*

So far we have discussed only the fate of dead plant remains and organic substances excreted by roots. Parts of many plants are eaten by herbivorous or omnivorous animals, including man, instead of decomposing in the soil. Part of the carbon in the food of animals is lost during their respiration and part is returned to the soil in excreta and dead bodies. Some may be eaten by animals, including man, that feed on the herbivores. In this way some of the carbon fixed by plants may pass through several animals before being finally released as carbon dioxide.

4.2.3 *Fossil fuels*

An important source of atmospheric carbon dioxide particularly during recent times in urban areas is the burning of fossil fuels, coal, oil and petroleum. Burning these fuels releases large quantities of carbon that was withdrawn from circulation by plant photosynthesis thousands of years ago. It has been calculated that if the world consumption of fuel continues to increase at the present rate, the average carbon dioxide content of the atmosphere will have increased to 400 p.p.m. by the year 2000. This could result in an increase of 20% in the rate of dry matter production by crops.

4.2.4 *Autotrophic bacteria*

Autotrophic bacteria can use the carbon in carbon dioxide directly to synthesize organic compounds, using energy from light or from oxidizing inorganic substances. A small part of this carbon may be returned in respiration, but most will not re-enter the cycle until the bacteria die or are eaten by predators.

A study of the diagram of the carbon cycle shows how much of the cycle is in soil, and to what extent it depends on the activities of soil organisms. Without these organisms, organic debris would accumulate, atmospheric carbon dioxide become depleted, and plants would eventually stop growing.

4.3 The nitrogen cycle

The circulation of nitrogen and carbon in nature (Fig. 4–2) have some features in common, but there are important differences. Organisms nearly always live in an environment containing much more nitrogen than carbon, but it is in a form that few organisms can use. Most plants use either ammonium or nitrate ions as their main source of nitrogen but can also use some organic nitrogen compounds, such as urea and amino acids. Nitrite cannot be used, and it is toxic. Like carbon, nitrogen is excreted as a constituent of organic compounds into the soil by roots, and also returned in plant debris, but plants lose little or none as gas.

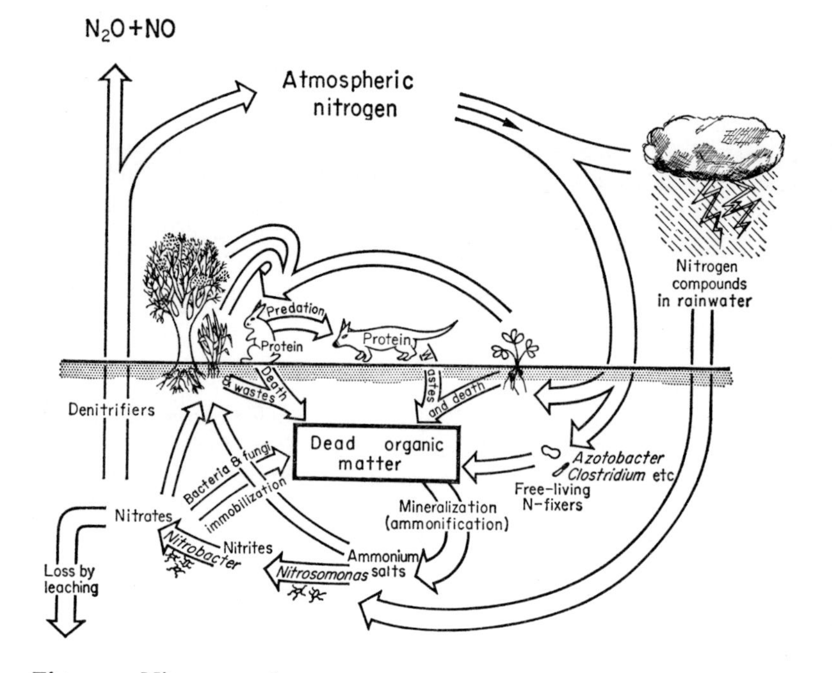

Fig. 4–2 Nitrogen cycle.

4·3·1 *Mineralization*

Soil bacteria are responsible for two main processes that convert organic nitrogen in the soil to a form that can be readily used by plants. The first is the changing of organic nitrogenous compounds to an inorganic or mineral form, ammonia or ammonium salts. Many types of soil bacteria, actinomycetes and fungi are responsible for this process of mineralization, which is in some respects analogous to the respiration of organic compounds with the release of carbon dioxide. It is a method by which micro-organisms growing on a nitrogen-rich substrate, such as protein or nucleic acid, get rid of excess nitrogen. During this process there may be a loss of gaseous ammonia to the atmosphere, but this is likely only in alkaline soils to which large quantities of manure or nitrogenous fertilizers have been added, or during decomposition of large masses of nitrogen-rich material such as farmyard manure. Nitrogen mineralization with the release of ammonia or formation of ammonium salts is often called ammonification, and the organisms responsible, the ammonifiers.

4.3.2 *Nitrification*

The second process is nitrification, the conversion of ammonium salts to nitrites and nitrates. Nitrogen is mostly taken up by plants as nitrate, and for this reason the process whereby nitrates are formed in soil is very important. The organisms chiefly responsible, the nitrifying bacteria, are autotrophic (2.1.2) and get all their energy by oxidizing ammonium to nitrite and nitrite to nitrate. Some of the energy released is used to obtain carbon by the reduction of carbon dioxide or bicarbonates. This provides the raw material for all the organic compounds they require. The efficiency of this process can be measured by the ratio of inorganic nitrogen oxidized to carbon dioxide carbon assimilated. For *Nitrosomonas europaea*, the best known species oxidizing ammonium to nitrite, the ratio is approximately 35:1, but for *Nitrobacter agilis*, which oxidizes nitrite to nitrate, it is between 76 and 135:1. Consequently nitrifying bacteria oxidize large quantities of ammonium or nitrite while making little growth.

Nitrification can also be brought about in culture by heterotrophic soil organisms, including bacteria, actinomycetes and fungi, but the significance of nitrification by these organisms in soil is thought to be small. Most can only oxidize ammonium to nitrite.

The first stage of nitrification is not immediately beneficial to plants, which can use ammonium salts but not nitrites which are harmful in more than very small concentrations. Nitrite is prevented from reaching toxic concentrations in soil by the second stage in the process, conversion of nitrite to nitrates. Fortunately, nitrite is usually oxidized faster than it is formed, but when large amounts of ammonium occurs in an alkaline soil, and the soil temperature is low, nitrite can accumulate.

Nitrate is a good source of nitrogen for plants, but is easily washed (or leached) out of the soil by rain. Thus nitrification can result in increased losses of nitrogen from the soil. A derivative of pyridine able selectively to inhibit nitrifying bacteria is marketed in America under the name of 'N-Serve'. In certain conditions, application of this compound to the soil can increase crop yield by reducing production of easily leached nitrate.

4.3.3 *Denitrification*

Denitrification is the process whereby nitrate is reduced to gaseous nitrogen or oxides of nitrogen by bacteria or fungi able to use nitrate as a source of oxygen. The process occurs in soil lacking oxygen, that is, in anaerobic or partially anaerobic conditions, and particularly when soil contains much organic matter. The process is agriculturally undesirable because it results in a net loss of nitrogen.

Organisms able to reduce nitrate to nitrogen or nitrous oxide include the bacteria *Pseudomonas denitrificans* and *Micrococcus denitrificans* and some fungi. Losses of nitrogen resulting from denitrification are greatest in badly drained soils and can be reduced by draining and ploughing to improve soil aeration.

4·3·4 *Immobilization*

Nitrogen may be temporarily lost to plants by another process called immobilization. When crop residues containing much carbon relative to nitrogen, i.e. having a high C:N ratio, are added to soil, microorganisms attacking such material will demand more nitrogen than it contains. To satisfy their needs they will quickly use all the mineral nitrogen present as ammonium or nitrate ions in the soil, and so make it unavailable to plants that cannot compete for it successfully. The nitrogen immobilized in this way is not lost from the soil, but is locked up in the cells of the microorganisms. As the amount of carbon in soil organic matter decreases, because carbon dioxide is lost by microbial respiration, the C:N ratio slowly decreases; the nitrogen is then gradually mineralized again and so becomes available to plants. When carbon-rich plant residues such as straw are ploughed into the soil it is necessary to apply nitrogenous fertilizers to compensate for the nitrogen immobilized. The microorganisms can then have their nitrogen without affecting crop yield.

4·3·5 *Symbiotic nitrogen fixation*

None of the processes described so far explain how combined nitrogen is added to soil. Were there no replenishment of soil nitrogen serious shortage would soon result because of losses by denitrification and leaching of soluble nitrogen compounds, particularly nitrates, out of the soil and to depths where they could not be reached by plant roots. Small amounts of nitrogen, in the form of ammonium and nitrate ions, occur in rain water, but these add only about 2 to 20 lb per acre per year; too little to replace losses. Most of the nitrogenous compounds in soil, on which plant growth depends, have originated by the biological fixation of atmospheric nitrogen.

Nitrogen fixation is brought about by two types of organisms, those that live in symbiotic association with higher plants, and those that are non-symbiotic or free-living. Members of the bacterial genus *Rhizobium* are the most important symbiotic nitrogen fixers. These bacteria, the root-nodule bacteria, fix nitrogen only when they are living in the roots of plants belonging to the family Leguminosae. This family includes such plants as clover, lucerne, peas, beans and vetches. The importance of nitrogen fixation by legumes in agriculture is illustrated by the fact that in New Zealand, where conditions are very favourable for clover growth and nitrogen fixation, nitrogenous fertilizers are rarely used on pastures, for as much as 500 lb of nitrogen can be fixed in each acre of pasture every year. In Britain nitrogen fixation is also important in pastures containing legumes, but nitrogenous fertilizers usually have to be applied to get maximum yields.

The nodule bacteria can live in soil away from their host plants, but gradually disappear when leguminous plants are not grown. There is some specificity in the symbiosis between *Rhizobium* and legumes, for individual strains of *Rhizobium* infect only certain legumes. For example, the bacteria that infect clover plants do not infect lucerne, and vice-versa. *Rhizobium* multiplies enormously in the soil near the roots of a suitable host plant (Plate 12), some of the bacteria penetrate root hairs, and at points on the roots the characteristic swellings, known as nodules, begin to develop (Plate 13). Inside the nodules the bacteria again multiply and become enlarged, and nitrogen is fixed. Nodules of a vigorously growing legume, when cut open, are usually seen to be bright red, because they contain a type of haemoglobin, and the colour indicates that the nodules have been fixing nitrogen. Some strains of *Rhizobium*, called ineffective, fix nitrogen either feebly or not at all. They form nodules that contain little or no red pigment.

Legume seed is often inoculated with the correct *Rhizobium* strain before sowing because of the limited period for which *Rhizobium* can survive in soil in the absence of its appropriate host plant. In Britain most lucerne seed is inoculated, because the lucerne *Rhizobium* is particularly short-lived, and without inoculation, nodulation and growth of lucerne is often poor.

For farmers, the association between legumes and *Rhizobium* is the only one of any significance in nitrogen fixation, but there are other symbiotic associations able to fix nitrogen. Non-legumes occurring in Britain in which root-nodule inhabiting organisms fix nitrogen, include Alder, Sea Buckthorn and Bog Myrtle. Although the subject of intensive study, the identity of the organisms in the nodules of these plants is still not certain. The evidence points to their being obligately—symbiotic actinomycetes. In some tropical and south temperate countries, coniferous trees belonging to the Podocarpaceae form large tracts of forest. The roots of these trees have nodules that usually contain fungi of the *Endogone* type. Recent work with isotopic nitrogen has shown that nitrogen is fixed in these nodules.

4.3.6 *Non-symbiotic nitrogen fixation*

Since the early days of soil bacteriology there has always been interest in the free-living or non-symbiotic nitrogen-fixing organisms. Many scientists have doubted whether these organisms provide an important source of nitrogen compounds for plants, but some recent work in Nigeria shows how important they can be. One experiment showed soil with a tropical grass and no legumes gained 100-130 lb of nitrogen per acre per year, compared with a gain of 100–200 lb in a similar soil containing legumes. The gain of nitrogen in the legume-free soil was due to free-living nitrogen fixers.

The best known of the free-living nitrogen fixing organisms are aerobic bacteria belonging to the genus *Azotobacter*. They occur in most fertile soils, that are not particularly acid or lacking in phosphate, but usually they are not very abundant. In desert and sand dune soils very large numbers of *Azotobacter* have sometimes been found in the root zone of plants. In these soils, which contain little nitrogen, the *Azotobacter* may provide a valuable addition to the plant's nitrogen supply. In tropical soils *Azotobacter* is often absent but in its place there are often bacteria belonging to a rather similar, but more acid-tolerant genus, *Beijerinckia*.

Equally good at fixing nitrogen, but often more abundant than *Azotobacter*, even in well-aerated soils, are bacteria belonging to the genus of anaerobic spore-forming bacteria, *Clostridium*.

Recent experiments using ^{15}N show that many soil organisms can fix small quantities of atmospheric nitrogen. Since some of these, such as species of *Aerobacter* and *Pseudomonas*, are often abundant in soil, it is quite possible that the total amount of nitrogen they fix may be as great as that fixed by the better known *Azotobacter* and *Clostridium*.

Several blue-green algae are very good nitrogen fixers. In suitable conditions, for example on the surface of bare, moist earth in the tropics, they may add valuable amounts of nitrogen to the soil. Unlike the other nitrogen fixers described, they can photosynthesize and so need no external supply of carbon compounds. In the tropics where paddy rice is grown, the whole nitrogen requirements of the rice crop may be provided by blue-green algae.

4·4 The phosphorus cycle (Fig. 4–3)

Phosphorus is an essential constituent of several important compounds always present in plants and animals. These include nucleic acids, nucleoproteins, phospholipids and phosphorylated sugars. Soil phosphorus occurs both in inorganic form as mineral salts, and in organic compounds. Most plants obtain the phosphorus they require from the soil only as orthophosphate ions (H_2PO_4), and these may be quite insufficient for healthy plant growth even though there is plenty of phosphorus in the soil in unavailable forms. Plants with mycorrhiza may be able to use some of this otherwise unavailable phosphorus. Most of the organic phosphorus in soil comes from plant and animal residues, but some also from the bodies of microorganisms that use orthophosphate and convert it into organic forms in their cells. These microorganisms are important because they compete directly with plants for available phosphorus; they are then said to immobilize phosphorus.

Not only does much of the phosphorus in soils occur naturally in forms that plants cannot use, but when phosphatic fertilizers, such as superphosphate, are added to the soil to improve the supply of phosphorus to the plant, much of this may be rapidly converted to unavailable inorganic compounds by a process called chemical fixation. Fortunately many bacteria and fungi can convert the phosphorus present in these compounds back into orthophosphate, by means of the organic acids they produce and possibly by other mechanisms. This action, sometimes called solubilization, is particularly active near roots where sugars from root exudates are

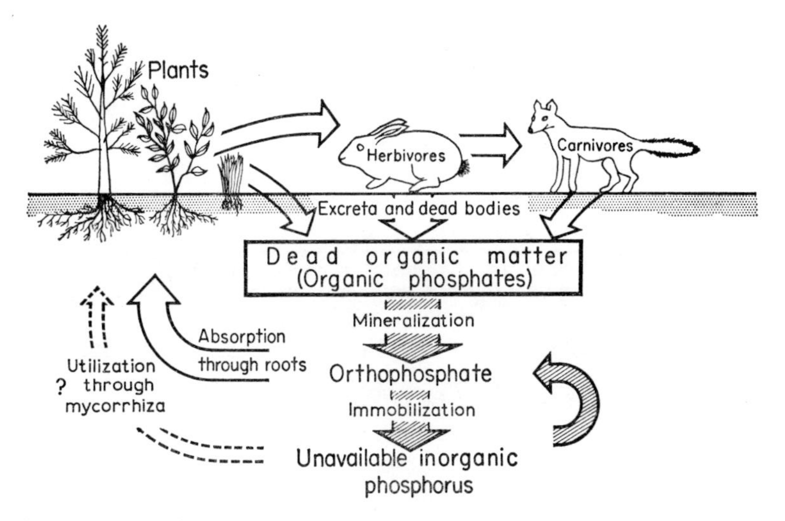

Fig. 4–3 Phosphorus cycle.

converted by microbial action into organic acids. Other microorganisms have enzymes able to attack many of the organic phosphorus compounds in the soil and release inorganic phosphate. This process is comparable to the mineralization of organic nitrogen compounds.

4·5 The sulphur cycle

Like nitrogen, sulphur is also an essential part of all living matter. In the same way as nitrogen, sulphur accumulates in the soil mainly as a constituent of organic compounds and does not become readily available to plants until it has been converted to sulphates, by the action of microorganisms. Plants can absorb the sulphur-containing amino acids, cystine, cystein and methionine, but these provide only a small proportion of their total requirements for sulphur. Under anaerobic soil conditions, the sulphur in organic compounds is converted first to H_2S and sulphides. These must be oxidized to sulphates before they can be used by plants. Two kinds of bacteria are mainly responsible for this process. One group are autotrophs belonging to the genus *Thiobacillus*, and are analogous to the autotrophic nitrifiers, deriving their energy from the oxidation of sulphides instead of ammonia or nitrites. The most studied species, *Thiobacillus thioxidans*, produces an acid environment under anaerobic conditions, and can tolerate a very low pH. The other important group of sulphide oxidizers are heterotrophic bacteria, fungi and actinomycetes, and in many soils they may be more important than *Thiobacillus*.

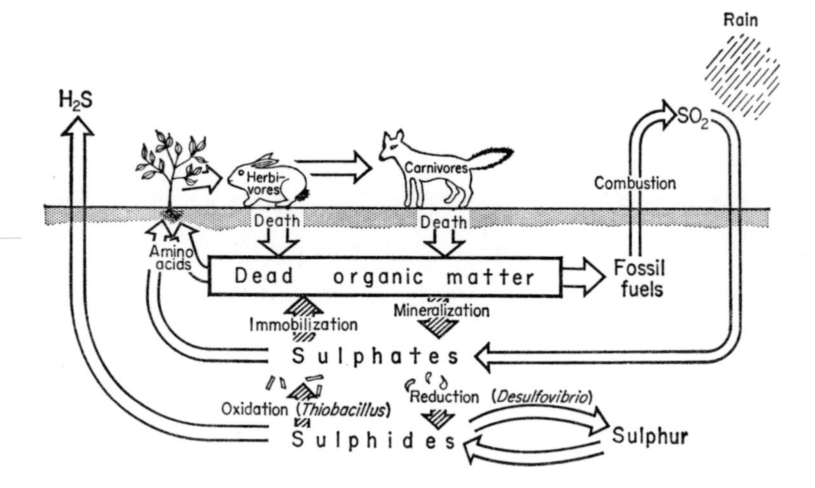

Fig. 4–4 Sulphur cycle.

The counterparts of the denitrifiers in the sulphur cycle are the sulphate-reducing bacteria. These are active principally under anaerobic conditions and can rapidly reduce sulphate and release H_2S. The best known sulphate-reducing species is *Desulfovibrio desulfuricans*, but various other anaerobic bacteria can also reduce sulphate.

4.6 The rhizosphere

We have now seen how soil organisms play an indispensable part in the circulation of nutrients, decomposing organic wastes, and transforming nutrients into forms that plants can use. These processes are going on throughout the *A* and *B* horizons of the soil, but are most active close to live plant roots. This region in which the root influences the activity of soil organisms is the rhizosphere. From the time the young root emerges it starts secreting many organic compounds. Sugars and amino acids predominate in the first week or two, but the exact composition of the root exudates depends both on the type of plant and on the conditions in which it is growing. As the organic substances from the root diffuse into the soil, fungal spores germinate and mycelium grows, much of it on the surface of the root. At the same time resting bacterial cells start to divide and increase greatly in numbers. Bacteria, actinomycetes and fungi are always many times more numerous in the rhizosphere than in soil away from roots, and there are also marked qualitative differences. For example, bacteria belonging to the genus *Pseudomonas* are strongly stimulated in the rhizosphere, whereas *Bacillus* spp. may not be much more numerous there than in non-rhizosphere soils. Multiplication of *Rhizobium* near the roots of leguminous plants is a special example of this rhizosphore effect.

The rhizosphere effect is strongest at the root surface and declines with increasing distance from it. Most of the organisms stimulated near the root are saphrophytes, but some are facultative parasites and in time penetrate the root cortex. There is intense competition for nutrients between the organisms in the rhizosphere. The most successful competitors are likely to be those that can grow fastest, can produce antibiotics and can resist antibiotics produced by other organisms. It is easy to show, for example, that bacteria resistant to the antibiotic streptomycin are much more numerous in the rhizosphere than away from roots. Some of the micro-organisms in the rhizosphere compete with the plant for nitrogen and phosphorus. Others benefit the plant by increasing the availability of nutrients or by synthesizing growth substances such as gibberellins, which are absorbed by the roots, and some micro-organisms, as we have seen, may become symbiotic partners of the plant.

4.7 Conclusion

It is hoped that this short account of the organisms that form an integral part of soil, and of their activities in soil formation and plant growth, will increase awareness of the importance and fascination of soil biology. Intensive study of nearly any non-aquatic ecological system reveals how biological processes in soil make their essential contribution, particularly in maintaining nutrient and energy flows.

It is important to realize that we are only beginning to understand some of the details of the complex interactions between soil organisms themselves, and between soil organisms and plants. A great deal more knowledge is needed before we will be able to manipulate soil organisms to permit the optimum growth of disease-free crops. There is no doubt that the problems of life in the soil present an exciting and formidable challenge to the biologist.

Bibliography

ALEXANDER, M. (1961) *Soil Microbiology*. John Wiley & Sons Ltd., New York & London.

BURGES, N. A. (1958) *Microorganisms in the Soil*. Hutchinson & Co. (Publishers) Ltd., London.

CLOUDSLEY-THOMPSON, J. L. & SANKEY, J. (1961) *Land Invertebrates*. Methuen & Co., Ltd., London.

COMBER, N. M. (revised by W. N. TOWNSEND) (1961) *An Introduction to the Scientific Study of Soil*. 4th ed. Edward Arnold (Publishers) Ltd., London.

ELLIS, A. E. (1926) *British Snails*. The University Press, Oxford.

GARRETT, S. D. (1963) *Soil Fungi and Soil Fertility*. Pergamon Press Ltd., Oxford.

GERARD, B. M. (1964) *A Synopsis of the British Lumbricidae*. Linnean Society of London Synopses of the British Fauna, London.

GOODEY, J. B. (1963) Laboratory Methods for work with Plant and Soil Nematodes, 4th ed. *Min. Agric. Fish. & Food Tech. Bull. no. 2*. H.M.S.O. London.

HARLEY, J. L. (1969) *The Biology of Mycorrhiza*, 2nd ed. Leonard Hill, London.

JACKS, G. V. (1954) *Soil*. Nelson (Thomas) & Sons, Ltd., Edinburgh.

JANUS, H. (1965) *The Young Specialist looks at Land and Freshwater Molluscs*, English edn. Burke Publishing Co., Ltd., London.

JOHNSON, L. F., CURL, E. A., BOND, J. H. & FRIBOURG, H. A. (1959) *Methods for Studying Soil Microflora—Disease Relationships*. Burgess Publishing Co., Minneapolis.

JONES, P. C. T. & MOLLISON, J. E. (1948) A Technique for the Quantitative Estimation of Soil Micro-organisms. *J. gen. Microbiol.* **2**; 54–69.

KEVAN, D. K. MCE. (1962) *Soil Animals*. Witherby (H. F. & Co.), Ltd., London.

KEVAN, D. K. MCE., ed. (1955) *Soil Zoology*. Butterworth & Co. (Publishers), Ltd., London.

QUICK, H. E. (1960) British Slugs. *Bull. Br. Mus. nat. Hist.* **6**, no. 3.

RUSSELL, Sir E. J. (1957) *The World of the Soil*. Collins (William), Sons & Co., Ltd., London.

RUSSELL, E. W. (1961) *Soil Conditions and Plant Growth*, 9th ed. Longmans, Green & Co., Ltd., London

SINGH, B. N. (1946) A Method of Estimating Numbers of Soil Protozoa, especially Amoebae, based on their differential feeding on bacteria. *Ann. appl. Biol.* **33**, 112–119.

SOUTHEY, J. F., ed. (1965) Plant Nematology, 2nd ed. *Min. Agric. Fish. & Food Tech. Bull. no.* 7. H.M.S.O. London.

STANIER, R. Y., DOUDOROFF, M. & ADELBERG, E. A. (1958) *General Microbiology*. Macmillan & Co., Ltd., London.